Wittgenstein's Lectures
Cambridge, 1930–1932

Also in this series:

Wittgenstein's Lectures
Cambridge, 1932–1935

From the Notes of Alice Ambrose and Margaret Macdonald

Edited by Alice Ambrose

Wittgenstein's Lectures

Cambridge, 1930–1932

From the Notes of
John King and Desmond Lee

Edited by Desmond Lee

The University of Chicago Press

This Phoenix edition is published by arrangement with Rowman and Littlefield, a division of Littlefield, Adams & Company.

The University of Chicago Press, Chicago 60637

89 88 87 86 85 84 83 82   1 2 3 4 5

Library of Congress Cataloging in Publication Data

Wittgenstein, Ludwig, 1889–1951.
  Wittgenstein's Lectures: Cambridge, 1930–1932.

  Reprint. Originally published: Totowa, N.J.: Rowman and Littlefield, 1980.
  Includes bibliographical references and index.
    1. Philosophy—Addresses, essays, lectures. 2. Mathematics—Philosophy—Addresses, essays, lectures. I. King, John, 1909–.
II. Lee, Henry Desmond Pritchard, Sir, 1908–. III. Title.
IV. Title: Lectures, Cambridge, 1930–1932.

[B3376.W561 1982]   192   81–19823
ISBN 0–226–90438–5 (pbk.)   AACR2

Contents

Acknowledgement

Our thanks are due to Mr. Brian McGuinness of The Queen's College, Oxford. It was he who suggested the publication of these notes, and he has been generous with his advice and help throughout.

Introduction

The lectures and discussions recorded here are in the main based on notes taken by myself (H.D.P.L.) and John King (J.E.K.) in the period from January 1930 to May 1932. Wittgenstein had returned to Cambridge early in 1929. I met him soon after his return, and used to see him fairly regularly until I went down in the summer of 1931; and when he started to lecture in the Lent Term of 1930 I attended his lectures for the Lent and Easter terms of that academic year and for the three terms of the following academic year.[1] J.E.K. first attended the lectures in October 1930, and his notes cover not only that academic year but also the academic year 1931–2. But J.E.K. has also secured notes taken by R.D. Townsend for part of the academic year 1930–31, and some rather briefer notes taken in the Lent and Easter terms of 1931 by John Inman. He and I are grateful to both of them for their generosity in allowing us to use their notes at our discretion to confirm and amplify our own.

The actual work of editing has fallen to me. For purposes of reference I have grouped the lectures and discussions into three series.

[1]The Cambridge academic year runs from October to June, and is divided into three terms, Michaelmas (October to December), Lent (January to March) and Easter (April to June).

ix

Series A. Lent and Easter terms 1930.

Series B. Michaelmas, Lent and Easter terms 1930-1931.

Series C. Academic year 1931–1932.

The sources for the three series are:

Series A. H.D.P.L.'s notes only

Series B. Michaelmas and Lent terms 1930, 1931.

    H.D.P.L.'s notes

    J.E.K.'s notes

    R.D.T.'s notes for Michaelmas term and Lent term to Lecture B X

    J.I.'s notes for Lent term

    Easter term 1931.

        H.D.P.L.'s notes

        J.I.'s notes

Series C. J.E.K.'s notes only

The lectures in series A and for the first two terms of Series B are numbered consecutively for each series. Because of the change in W.'s manner of lecturing, referred to below, there are no lecture numbers for the Easter term 1931 or for Series C.

When there was more than one source at my disposal I have based my version on my own notes and J.E.K.'s. His method was, so far as possible, to take down what W. said verbatim. My own notes, though again, so far as I remember, they are generally in W.'s words, are more selective, and more concerned to bring out the structure of the argument than with verbal detail. But since it did not seem sensible to present readers with problems of source criticism, and since the notes fitted well together into a whole, I have produced (with J.E.K.'s agreement and assistance) a single version. In doing so I have followed as closely as possible the wording of the notes as they stand. I have made some changes to make the two versions fit together, some to make the language grammatical or to expand the brevity of the note form, and I have omitted what seemed unintelligible. R.D.T.'s and J.I.'s notes have been a most valuable supplement. Not only have they provided confirmation of the topics dealt with, but also, just as my own and J.E.K.'s notes often supplement each other, they have been a source both of amplification and of particular phrases and examples. The traffic lights in Lecture B VII for example occur in R.D.T.'s notes (and are confirmed by a reference in Moore's articles[2]) and his notes being the fuller have been particularly valuable.

The numbered lectures in Series A and B are further divided into numbered paragraphs. These numbers are not based on any formal divisions made by W. when lecturing. His lectures may, I suspect, have been more carefully prepared than they appeared at first sight. He seemed to take up a topic or argument, perhaps following up something from a previous lecture or discussion, and to pursue the argument wherever it led, with frequent digressions and asides, and without any set plan. Yet the lectures have an overall coherence and shape, despite digressions and repetition, and if the numbered paragraphs are an importation, J.E.K. and I both feel that they serve, if only for reference purposes, to group topics together, and they do correspond, if only roughly, to paragraphings and groupings in our notes. They were not however a feature of the lectures as delivered, which had little formal shape, were liable to lapse into discussion, and were full of refinements and clarifications, which gave the impression that he was thinking them out as he spoke to us.

W.'s method of lecturing up to the end of the Lent term 1931 was to give each week one lecture of an hour's duration followed later in the week by a two hour discussion. The

---

[2]*Mind*, Jan 1954 p. 13.

lecture took place in the Arts School lecture room block, the discussion in rooms lent by R.E. Priestley (later Sir Raymond Priestley) then Secretary General of the Faculties. They were attended by a mixture of undergraduates and graduates, Professor Moore in particular being regularly present.

This arrangement, with its combination of formal lecture (though as I have said W.'s technique was anything but formal) and less formal discussion suited me personally very well. Though some of the material in my Miscellaneous Notes (see below) may have come from the discussions, I relied on the whole on the lectures for the exposition of his thought and regarded the discussions as explanatory and supplementary and took no regular formal notes. But at the beginning of the Easter term 1931 W. decided on a change. I have no clear recollection myself, but J.E.K. writes "I well remember that W. suddenly broke off his first lecture and turning to Professor Moore enquired whether it would be in order and acceptable to the authorities if in future we met in his rooms in Whewell's Court instead of in the Arts School. He had never liked the formality of the lecture room, and his manner and style were more suited to a more intimate and less conventional approach. Moore said he saw no objection to such a move, and thereafter we met in W.'s rooms at the top of one of the staircases in Whewell's Court." J.E.K. goes on to remark that the change is evidenced by the different sets of notes for that term. "I took none—on the assumption that it was too difficult to take notes on one's knee sitting on a deck chair; Townsend's are a blank; Inman's brief and incomplete." He refers to mine as "lengthy", but they are not so full as those for earlier terms and there is no indication of breaks between lectures. Note-taking was not so easy, and lecture and discussion tended to merge.

J.E.K. did however resume note-taking in the academic

year 1931–32. For that year his are the only notes available to me, as the others, like myself, left Cambridge in the summer of 1931. The editing of his notes, with no others for check or comparison, has not been easy. It has not been possible to identify breaks between sessions, and the present grouping by topics, though there is some indication of it in the notes, is largely due to J.E.K. and myself. One particular group consists of comments on views expressed by Dr (later Professor) Broad. I have been able to check Broad's views with my own notes on his course on the Elements of Philosophy to which W.'s comments clearly refer. B. lectured from a typescript from which he read very slowly, with frequent repetition, and my notes are therefore pretty well verbatim. W.'s comments do not all occur consecutively in J.E.K.'s notes, but I have for convenience grouped them together. Otherwise the order of topics in series C follows the order of J.E.K.'s notes.[3]

Conditions for note-taking were not easy, and J.E.K. must again speak for himself. "Before the 1931–2 session I decided that I could take down notes on my knee at these discussion classes. I equipped myself with a typist's note-book, which I used at the discussion classes; but bought also a larger stiff-backed book into which I copied what I had written with difficulty in Whewell's Court.

"W.'s room was square, with the window on the left side of the wall which faced you as you entered. He sat near the window, with the light coming over his left shoulder, at a small collapsible card-table, on which there was a large ledger-like book which he used for his own writing. A

---

[3]J.E.K. has some doubts whether these comments were made in the course of the 1931–32 lectures/discussions, and thinks they may come from some other source (possibly M. O'C. Drury). But they are included in his Wittgenstein notebook, are reminiscent of other passages there, and he remembers some of the illustrations as having been used by Wittgenstein himself. I have therefore included them.

number of chairs and deck-chairs were brought into this room and arranged in a semi-circle for those attending. There was a blackboard on W.'s left. The numbers attending were about ten or fifteen, including Moore, who would sit huddled in his chair smoking his pipe, which he was continuously lighting and relighting. Some other dons attended occasionally.

"To the best of my ability I concentrated on taking down whatever W. said verbatim. I never made any attempt to find my own terms, comparisons or examples, nor to alter his words or their order. The effort of note-taking made such changes impossible, even if I had felt capable of making them. W. never dictated notes but I treated his lectures and discussions as if he were doing so. Of course not everything could be got down, but I got down all I could." The difficulty lay in following what was often a difficult argument, with frequent digressions, harking back and repetition, and not in any lack of command of English. W. had a complete command of the language. If an occasional Germanic way of expression crept in it was hardly noticeable; and if he would often hesitate and pause before speaking it was in J.E.K.'s words from "his intense desire to pick just the right word or phrase for his purpose, or to choose the most telling illustration or example to convey his meaning. He must have the exact word or phrase; nothing else would do."

In editing Series C I have been in constant communication with J.E.K., and though some passages have had to be eliminated as unintelligible, the series contains the bulk of his notes for this period.

I used also to keep, at the back of my own lecture notebook, a record of things said by W. in the course of various discussions. Some of these discussions were quite informal and individual; some were the more formal discussions after his lectures (referred to above); some were a series of individual discussions which I had with him during the academic year 1930–31. This record is reproduced here under the heading Miscellaneous Notes. These notes are inevitably disjointed, without pattern or plan, but I hope of sufficient interest to justify their inclusion. The first entry among them was a note of the lecture on Ethics which he gave to the Heretics society in Cambridge on November 17th 1929. The lecture is now published in full in *Philosophical Review* (Jan. 1965) and my own note accordingly not reproduced here.

In a few places it has been possible to get some help from published works of W. which date, roughly, to this period of his thinking: *Philosophical Remarks, Philosophical Grammar,* and *The Blue and Brown Books.* But I have deliberately refrained from using them in the actual process of editing, as it seemed to me that the use of other sources would detract from any value the publication of these notes might have. I have therefore used them only in a few places in order to confirm or clarify our original notes (e.g. the French politician in Lecture B III occurs both in *Ph.R.* and in *Ph.G.*), but have not otherwise drawn upon them.

We hope that anyone who reads this book will remember its necessary limitations. The aim has been to reproduce in intelligible form lecture notes taken by undergraduates at a time when the ideas and methods it records were new and unfamiliar. The original notes can hardly be free from error and incompleteness; nor can their reproduction at this distance of time. But when the development of W.'s thought is discussed they may produce evidence of what he was in fact saying at the time when they were taken and, perhaps more relevant, of what he thought it important to say within the limited compass allowed by a course of university lectures and discussions.

DESMOND LEE

Wittgenstein's Lectures
Table of dates for Series A. and B.

The lectures in Series A. were given on Monday at 5.0 pm, the discussions took place on Thursday from 5.0–7.0 pm. The lectures in Series B. were given at noon on Monday and the discussion took place on Friday from 5.0 to 7.0 pm.

The dates (including the dates of the two discussions referred to on pp. 5, 14) were as follows:

Lent Term 1930

A I	January 20th
A II	January 27th
Discussion	January 30th
A III	February 2nd
A IV (Lecture and discussion)	
	February 13th
A V	February 17th
A VI	February 24th
A VII	March 3rd
A VIII	March 10th
Discussion	March 13th

Easter Term 1930

A IX	April 28th
A X	May 5th
A XI	May 19th

Michaelmas Term 1930

B I	October 13th
B II	October 20th
B III	October 27th
B IV	November 3rd
B V	November 10th
B VI	November 17th

B VII   December 1st

(There was no lecture on November 24th as Wittgenstein had flu.)

Lent Term 1931

B VIII	January 19th
B IX	January 26th
B X	February 2nd
B XI	February 9th
B XII	February 16th
B XIII	February 23rd
B XIV	March 2nd
B XV	March 9th

Easter Term 1931

At the beginning of this term the change referred to in the Introduction (p. xii) took place and it is no longer possible to distinguish between sessions, but they appear to have run from April 27th to June 1st.

ὁ μὲν ἐντὸς τῆς ψυχῆς πρὸς αὐτὴν διάλογος ἄνευ φωνῆς
γιγνόμενος τοῦτ᾽ αὐτὸ ἡμῖν ἐπωνομάσθη διάνοια·

Plato, *Sophist* 263e

SERIES A: 1930

Lent Term 1930

Lecture A I

1. Philosophy is the attempt to be rid of a particular kind of puzzlement. This "philosophic" puzzlement is one of the intellect and not of instinct. Philosophic puzzles are irrelevant to our every-day life. They are puzzles of *language.* Instinctively we use language rightly; but to the intellect this use is a puzzle.

2. Language consists of propositions (excluding for the moment so-called mathematical propositions). A proposition is a picture of reality, and we compare proposition with reality. We give prescriptions for action in propositions, and these prescriptions must have some picture-pictured relation with reality. The prescriptions we give (the signals, symbols which we use) must have a *general,* arranged significance and must be interpreted in *particular* instances: e.g. there is a *general* arrangement that whenever a railway signal stands in a certain position a train must stop and this is interpreted in a *particular* case. Thus language can convey something new, and we can interpret the general arrangements of language in particular instances.

3. A proposition must have the right multiplicity: for example, a command must have the same multiplicity as the action which it commands or prescribes. Thus a command to go from x to y along the lines in Figure 1 must give the right number of movements and turns to be made.

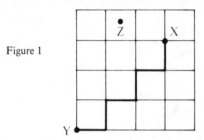

Figure 1

And if language can only describe a route which follows the lines, then it will be impossible to describe or prescribe a route to a point z which is not on the lines.

Language may be compared to the controls of a machine. These have the same multiplicity as the movements which the machine is capable of making. You can't get four speeds out of a three-speed gearbox. To try to do so would be the equivalent of talking nonsense in language.

The same requirements apply to a description as to a command or a prescription. A description is verified or falsified by comparison with reality, with which it may correspond, or not, and so be true or false. This is true of propositions generally.

4. The proposition, having multiplicity, is therefore a complex. Its constituents are words. Have words meaning apart from their occurrence in propositions? Words function only *in* propositions, like the levers in a machine. Apart from propositions they have no function, no meaning.

The constituents of propositions have been said to be subject and predicate, parts of speech, relations of some

kind. But this is incorrect. I could express some propositions, for example, in a series of raps. Thus the journey from
A to B in Figure 2 can be described in raps, if it is understood that one rap indicates a move of x to the right and two
raps a move of x upwards.

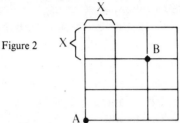

Figure 2

Substantives and other parts of speech are only essential in
our language. Such linguistic classifications are very misleading, as can be seen by substituting words for each other
in propositions of the same linguistic form. Substitution is
only possible when the words are of the same kind. Thus—

> This book is blue.
> This book is brown.
> I am tired.

Here, *brown* and *blue* can be substituted for each other, *tired*
for neither and neither for *tired*. Such substitution makes
nonsense because the words are not all of the same kind.
Blue and *brown* are of the same kind, for the substitution of
one for the other, though it may falsify the proposition, does
not make nonsense of it.

We can see this also by considering Lewis Carroll's
nonsense rhymes.

> 'Twas brillig and the slithy toves
> Did gyre and gimble in the wabe.

This can be analysed into subject and predicate and parts of
speech, but is nonsense.

This shows that such analyses of the constituents of propositions are not correct.

Lecture A II

1.   It will be objected that philosophy is concerned with thought, not language. In fact its concern is rather with the internal relations of thought, and these must be examined through the expressions conveying them. When a philosophical problem is elucidated, some confusion of expression is always exposed. For example—2 × 2 *is* four, the door *is* brown. (Remember the trouble that the word "is" has given to philosophers.) The confusion is resolved by writing = for *is* in the first phrase and ε is the second.

But to say that philosophy is concerned with problems of expression is not to minimise those problems.

2.   The proposition is a picture of reality.
Two senses of picture—
   (1)   A portrait, which is like, resembles, is similar to that of which it is a portrait.
   (2)   Something which is *intended* to be a picture of another without resembling it in sense (1). That it is a picture consists in the intention.

3.   What does it mean to "understand" a proposition? I may understand your command, but not obey it, I may think about it without acting on it. (A similar difficulty arises also in memory and expectation.)

How then do we discover that someone understands a proposition? If I want to show a man what I mean by "lifting my arm in 8 minutes time" I cannot do so by lifting my arm in 8 minutes time; I must e.g. lift my arm now. The time-gap is filled by expectation. Expectation does not contain its own

4

fulfilment; the fulfilment or non-fulfilment answers the expectation but is not contained in it, nor can expectation and fulfilment be set side by side. Yet I can say when my expectation is fulfilled (verified), and indeed in some cases how nearly it is fulfilled: e.g. how nearly alike are the colour expected and the colour actually seen.

> I expect a red patch.
> I see a red patch.

The two facts, the expectation and the actual seeing, have the same logical multiplicity, and it is in this logical multiplicity that expectation and event are comparable, not in the sense that portrait and original are.

4.   We learn/teach language by using it. The linguistic convention is conveyed by linking the proposition with its verification. To "understand" is to be led by linguistic convention to a right expectation; and of the expectation we can only say that it must have the same logical multiplicity as the event. A proposition must have the same logical multiplicity as the fact to which it refers.

To be of the same logical kind two words/expressions must be able to be substituted for each other.

> I see a) the moon.
>   b) the surface of the moon.

a) and b) seem to be the same kind of expression, but they cannot always be substituted for each other. We can talk about the area of the surface, but not of the moon itself.

5.[1]   The *thought* that x is the case is as different from x being the case as is the *proposition* "x is the case". Both

---

[1]The points in this paragraph were added in the subsequent discussion.

thought and proposition indicate the method of finding out whether x is the case and point to the space (visual, tactual, etc.) in which to look. For example, "The clock will strike in five minutes' time". Here (1) you have to wait for time to elapse (temporal space), (2) if the proposition is true you will hear the clock strike (auditory space).

A system of language can only express expectation if it can express the present state of affairs.

Lecture A III

1.   A proposition has the same kind of relation to reality that a measuring rod has to an object. This is not a simile; the measuring rod is an example of the relation.

Thus we could mark points c and d on a clock and a thermometer, and the two marks would say that when the hand of the clock reached c the mercury in the thermometer would stand at d. These marks are a picture of the positions of the clock and mercury at c and d. The marks must express possible positions of the mercury and the clock hand, and we must also be able to express, by differently placed marks, the present state of affairs.

Figure 3

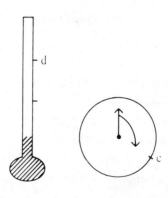

6

We must have an arrangement which tells us how to apply the measuring-rod, a method of application. The measuring-rod must have length, that is, be in the same space as the thing measured, and we must have made arrangements how to apply it.

These conditions apply to propositions also.

2.   The rules in which the method of application is expressed belong to, are part of language. If I say, "This desk is 3 ft. high," I must know to which desk I am referring and its existence is part of language. All the conditions necessary for the comparison of proposition and reality belong to, are part of the rules governing the application of language to reality. If F (3) means "Apply the measuring-rod x three times to object O and that will give you the height of O," then the existence of O and the measuring-rod are part of the proposition F (3). What is not part of it is the height of O.

3.   Thus, in this case, written symbols alone are not enough. And this is often so. If I want a wall painted a particular colour I cannot describe it in words but must give a specimen of the colour. This specimen is then part of the proposition. Similarly, a memory-image may be part of the symbol, the words alone not being enough. I may have a memory-image of a colour which is then part of the symbolism, as was the specimen colour. Memory-image or imagination is a picture in the sense that it has the same multiplicity as the fact or object remembered or imagined. Most propositions presuppose some sort of memory or imagination.

But we *can* express the element normally supplied by memory and imagination in the symbol (written or spoken) itself. This is what we do in the process of analysing a proposition; we get the imagination into the symbol and so complete its multiplicity.

4.   The multiplicity of language is given by grammar. A proposition must have the same multiplicity as the fact which it expresses: it must have the same degree of freedom. We must be able to do as much with language as can happen in fact. Grammar lets us do some things with language and not others; it fixes the degree of freedom.

The colour octahedron (Fig. 4) is used in psychology to represent the scheme of colours. But it is really a part of grammar, not of psychology. It tells us what we can do: we can speak of a greenish blue but not of a greenish red, etc.

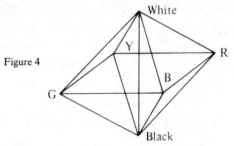

Figure 4

But grammar is not entirely a matter of arbitrary choice. It must enable us to express the multiplicity of facts, give us the same degree of freedom as do the facts.

Euclidean geometry is a part of grammar. It is a convention of expression and so part of grammar. (Minkowski accounts for the result of the Michelson-Morley experiment by a new geometry, Fitzgerald by contraction. These are merely two expressions of the same fact; we can adopt either, unless a decisive experiment is possible between them.)

We can now see what is meant when we say that something is possible if we can imagine it. If a thing has happened in the past, that does not prove that it is possible now (though we tend to think so). The possibility is expressed (contained) in language itself.

Lecture A IV[2]

1. Ogden & Richards and Russell[3] consider that the relation of proposition to fact is an external relation; this is not correct. It is an internal relation. (An internal relation cannot be otherwise; it is given in the terms involved, in the nature of proposition and fact.) On Russell's view you need a tertium quid besides the expectation and the fact fulfilling it; so if I expect x and x happens, something else is needed, e.g. something that happens in my head, to link expectation and fulfilment. But how do I know that it is the right something? Do I, on the same principle, need a fourth something? If so, we have an infinite regress, and I can never know that my expectation has been fulfilled. (We can always ask for a further description of any criterion given of meaning or fulfilment; which produces an infinite regress.)

Russell treats wish (expectation) and hunger as if they were on the same level. But several things will satisfy my hunger, my wish (expectation) can only be fulfilled by something definite.

2. We cannot say "A proposition p is possible". If p has sense, that in itself shows that it is possible. If p were not possible it would not be a proposition at all. All that we could mean by "p is possible" would be "p has sense"; there could be no indication whether what p asserts has happened or could happen. So "p is possible" is not a legitimate proposition. (Possibility here means logical possibility.)

3. True propositions describe reality. Grammar is a mirror of reality. Grammar enables us to express true and false

---

[2]No lecture; lecture and discussion combined.
[3]Ogden and Richards, *The Meaning of Meaning;* Russell, *Analysis of Mind,*

propositions; and that it does so tells us something about the world. What can be expressed about the world by grammar being what it is cannot be expressed in a proposition. For this proposition would presuppose its own truth, i.e. presuppose grammar.

Lecture A V

1.   Language represents in two ways—

(1) Its propositions represent a state of affairs and are either true or false.

(2) But in order that propositions may be able to represent at all something further is needed which is the same both in language and in reality. For example, a picture can represent a scene rightly or wrongly; but both in picture and scene pictured there will be colour and light and shade.

Thought must have the logical form of reality if it is to be thought at all.

Grammar is not the expression of what is the case but of what is possible. There is a sense therefore in which possibility is logical form.

2.   There are no logical concepts, such for example as "thing", "complex" or "number". Such terms are expressions for logical forms, not concepts.

Roughly speaking, a concept can be expressed as a propositional function: e.g. $\phi(\ \ ) = (\ \ )$ is a man. But we cannot say $\phi(\ \ ) = (\ \ )$ is a number. Such logical concepts are pseudo-concepts and cannot be predicated as ordinary concepts are. They are properly expressed by a variable together with the rules applying to it, the rules for obtaining its values. So I cannot write $(\exists x).x$ is a number or $(\exists x).x$ is a thing. If I use this notation I must write $(\exists x_{number}).\ \phi x$, meaning that there are certain variables to which specified rules apply: the pseudo-concept occurs inside the ∃ bracket, the true concept outside. All apparent logical concepts are to be expressed by a variable plus the grammatical rules governing its use.

(If it has sense to say "There are four primary colours" it must have sense to say "There are five primary colours".)

3.   That there are thus no logical concepts explains why it is nonsensical to classify in philosophy and logic. In philosophy and logic there are no "kinds" of thing, term and so on. In the sciences we draw distinctions and classify, and we do so by means of propositions which are true of one kind of thing and false of another. This is just what we can't do in logic.

Lecture A VI

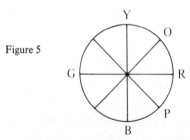

Figure 5

1.   In fig. 5 purple is between red and blue in quite a different sense from that in which red is between purple and orange. You can't have a mixture of orange and purple (colours not pigments). Hence it would be less misleading to use a square (fig. 6). But there is no middle point between red

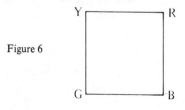

Figure 6

and blue (cf. points with no metric in geometry) and if the diagram suggests this, it is still misleading.

"Primary colour" and "colour" are pseudo-concepts. It is nonsense to say "Red is a colour," and to say "There are four primary colours" is the same as to say "There are red, blue, green and yellow." The pseudo-concept (colour) draws a boundary *of* language, the concept proper (red) draws a boundary *in* language.

Lecture A VII

1.   To all this it may be objected "But you are talking about colour (and similar terms) all the time." To this objection the answer is that what we are doing is giving the grammatical rules and conventions applying to colour, etc. To which it may be further objected, "Are you then talking of 'mere convention', of mere convention in the sense that the rules of chess or any other game are 'mere convention'?" Grammar certainly is not merely the conventions of a game in this sense, the game of language. What distinguishes language from a game in this sense is its application to reality. This application is not shown in grammar; the application of the signs is outside the signs, the picture does not contain its own application. Language is connected with reality by picturing it, but that connection cannot be made in language, explained by language. Try to make the connection in a proposition such as "This is green". The "this" in the proposition may refer to the colour of a panel, piece of wood or what not. But this does not tell us anything about the connection of green and reality: it presupposes that we already understand colour words. Someone might guess our meaning, but we have not made a connection. Alternatively we might say "This is green" pointing to a particular coloured patch. But

here our pointing is used as a sign of equivalence: This = green, giving a definition. But there is still no connection unless we already know what kind of word "green" is.

Conventions presuppose the applications of language: they do not talk about the application of language.

All grammar is a theory of logical types; and logical types do not talk about the application of language. Russell failed to see this.

Johnson says that the distinguishing characteristic of colours is their way of difference from each other. Red differs from green in a way that red does not differ from chalk. But how do you know this? "It is formally not experimentally verified" (W.E. Johnson, *Logic* Vol. I p. 56). But this is nonsense. It is as if you were to say you could tell if a portrait was like its original by looking at the portrait alone.

Language shows the possibility of constructing true or false propositions, but not the truth or falsehood of any particular proposition. So there are no true a priori propositions (mathematical propositions so called are not propositions at all).

2. *Infinity*

Infinite is not an answer to the question "How many", the infinite is not a number. It is an infinite possibility in language of constructing propositions. The word "all" refers to an extension; but it is impossible to refer to an infinite extension. Infinity is the property of a law, not of an extension.

Lecture A VIII

1. "Possibility" is what is represented by a proposition having sense (grammar is the expression of what is possible,

Lecture A VI above). An "infinite possibility" is not expressed by a proposition asserting it but by a law of construction. Infinite divisibility is not expressed by a proposition asserting that division has taken place, but by a law giving an infinite possibility of propositions asserting further division (division into a given number, not into an infinite number). To a proposition asserting a given number of divisions there is a corresponding reality: there is no infinite reality corresponding to an infinite possibility. Infinity is not a number but the property of a law.

The rule for infinity can be expressed symbolically as follows: $[f(1), f(\xi), f(\xi + 1)]$. Note that we have to go on step by step, starting from $f(1)$. This is not the kind of generality represented by $(x)\,\phi x$.

2. Infinite divisibility and space.

It has been maintained that space is not infinitely divisible. But experiment can prove nothing about possibility, and to show e.g. that there are minimal visible parts of visual space does not prove that space is not infinitely *divisible.*

Geometry is used in physics in a sense different from that in which it is part of grammar. The space of which physics talks is space in a different sense from space in the sense of visual or tactile space.

Discussion.
Senses of "all" and generality.

(i) $(x).\ \phi x$: for all values of x, ϕx; e.g. *a b c d* are values of x, being names of the four primary colours.

($(x).\ \phi x$ expresses a logical product, $(\exists x)\ \phi x$ a logical sum).

(ii) Incomplete (indefinite) description: I met a man.

(iii) Generality in mathematics, which is given by induction.

Easter Term 1930

Lecture A IX

1.   Different senses of generality have structural properties in common, not qualities; they don't fall under a genus as its species, e.g. different kinds of number have in common certain rules of grammar.

(i) (x). ϕx The values of x will be proper names (a proper name in this sense may be defined by saying that if substituted for ξ in "ξ exists" it makes nonsense). And if I write ϕx, $\phi \beta$, $\phi \gamma$, $\phi \delta$ where a, β, γ, δ are values of x (e.g. primary colours) I do not need another proposition to say "and these are all". For if I say "And there is no other" it must also make sense to say "There *is* another". ϕa, $\phi \beta$, $\phi \gamma$, $\phi \delta$ tells me everything.

(ii) I met a man. This cannot mean a disjunction—I met a or b or c. A disjunction is not what we mean. The generality here is of the same kind as that in the proposition There is a circle in this square. But if we write this $(\exists x)$. ϕx, i.e. there is something such that it is a circle in this square, this is equivalent to regarding "a circle in this square" as a possible predicate. But a predicate what of? What is the something that is a circle in the square? If there is something of which it is true to say that it is a circle in the square, it must also have sense to say that it is *not* a circle in the square. Could it be a predicate of the point which is the centre of the circle, and which can be determined by co-ordinates? Then we can say there are two co-ordinates which give the centre of the circle in this square. But it would also have to make sense to write $(\sim \exists x) \sim \phi x$ ("All circles are in the square"). This would be an infinite totality, which is nonsense.

15

2.   There is a different kind of generality which applies to hypotheses. A proposition can be verified; a hypothesis cannot, but is a law or rule for constructing propositions and looks to the future—i e. enables us to construct propositions which say what will occur and which can be verified or falsified.

Lecture A X

1.   There are no gaps in grammar; grammar is always complete. This seems paradoxical because we seem to be able to make *discoveries* in mathematics and logic. But we can only have incompleteness *in* a space. There is a fundamental difference between a grammatical "discovery" and any other type of discovery. It is a difference of logical type, and it is misleading to use the same word for both. So Sheffer's stroke-function was not a discovery, filling up a gap in grammar; Sheffer found a new space.

2.   When we use a hypothesis we can draw a conclusion on evidence from several sources. We cannot do this with propositions about sense-data nor with propositions of logic and mathematics. In propositions about immediate experience and in logic and mathematics there is no question of different sources of evidence; but in hypotheses there is.

3.   You cannot even suspect what you later prove by a method before you know the method. You might think that a man might, by measuring the angles of actual triangles, arrive experimentally at, or suspect from his experiments, what he later proved, namely that the internal angles of a triangle were equivalent to two right angles. But this is not so: what he *proves* is of an entirely different kind from what he arrives at or suspects as a result of experiments. Geometry does not prophesy, but says that if the measured

angles of a triangle add up to 181° there has been a mistake.

4.   You cannot look for a space, but only for things in a space. "Space" in this sense means everything of which you must be certain in order to be able to ask a question. What one looks for *in a space* must be completely describable; a description must be able to give me everything I subsequently find. "Description" is not in any sense "incomplete", as Russell assumes in his distinction between "acquaintance" and "description". He thinks that acquaintance gives you something more than description.

Logical discovery is quite different from finding something *in a space*; in logical discovery if we could describe completely what we were looking for we should already have it.

5.   That a generality has an infinity of special cases does not make it more complex than if it had only three or four special cases. It is true that a proposition asserts all that follows from it, and in a sense therefore to have four special cases as consequences is to be more complex than to have three. But a generality with an infinity of special cases is of an entirely different logical kind from this. It does not assert an infinite number of propositions. A Euclidean proof is not infinitely complex. The proof that the internal angles of a triangle total 180° is a proof about space and not about any particular triangle. Infinite possibility is always represented by an infinite possibility in the symbolism.

Lecture A XI

1.   Generality in geometry (sense (iii), p. 14). It appears at first sight as if there were concepts of points, lines, etc. and things that are points, lines, etc. This is not so. A point is not a concept at all. What then is generality in geometry? It has

two senses—(i) the generality of geometrical rules, (ii) the generality of the application of geometry. The application depends on what the world is like.

This generality of application is also the generality of arithmetic.

2. Proof in mathematics.

There are two quite different kinds of proof in mathematics.

The first proceeds by certain rules of substitution (axioms) from one equation to another.

The second is proof by mathematical induction.

As an example of the second take the Associative Law for the addition of numbers.

We have—

Symbols—numbers $1, (1)+1, ((1)+1)+1---$

or use $[1, \xi, \xi+1]$

Definition—$a+(b+1) = (a+b)+1$ Def.

The Associative Law $a+(b+c) = (a+b)+c$ to be proved by means of these.

We then write—

$a+(b+(c+1)) = a+((b+c)+1) = (a+(b+c)+1)$
$= ((a+b)+c)+1 = (a+b)+(c+1)$

The first and second equations follow in virtue of the Definition; so also does the fourth. But the third $((a+(b+c) +1) = ((a+b)+c)+1)$ appears to use the Associative Law which is what is to be proved.

$(b+c)$ in the second equation is not a number as b is in the definition. But we can substitute 1 for c, which puts this right. And if we can substitute 1 we can also substitute 2, and so on for all numbers.

Thus proof lies in a series of such chains of equations (substituting 1,2,3 and so on), or rather in the law by which you can formulate these chains. Each chain is a proof in sense (1) above, and the proof may be symbolised by a spiral—ℰ. What is important is the *law* by which we can construct such chains of equations. In what sense is this a proof? The proof shows that the Associative Law is applicable to numbers—i.e. that variables $a + (b + c) = (a + b) + c$ may be numbers.

From the associative follow other laws by strict proof. $A \rightarrow B \rightarrow C \rightarrow D$
A is proved by its spiral, B by its spiral and so on.
There is an internal relation between spirals answering to the internal relation between A and B.

$$A \dashrightarrow B \dashrightarrow C \dashrightarrow D$$
$$\uparrow \qquad \uparrow \qquad \uparrow \qquad \uparrow$$
$$ℰ \qquad ℰ \qquad ℰ \qquad ℰ$$

The relation between A and its spiral is not strict proof. A cannot be questioned and we cannot ask whether it is true or false; and the induction (the spiral) cannot be questioned. All that is questionable is their relations in a system.

3.   I might as well question the laws of logic as the laws of chess. If I change the rules it is a different game and there is an end of it.

Note on section 2.
References.   G.E. Moore, *Mind* Jan '55 p 5–7
Wittgenstein, *Philosophical Remarks* p 193 ff,
Philosophical Grammar p. 397 ff.

"quite different kinds of proof": J. E. K. in his notes for Easter Term 1932 has "If you want proof of this (the Associative Law) you expect something similar to $(a + b)^2 = a^2 + b^2 + 2ab$. But you don't get it." cf. Moore p 6 "the fallacy of supposing that ---- 'proof' always means the same," and *Ph. G.* p 398.

"appears to use the Associative Law." cf. Moore p 6: M. writes that though the proof seems to assume at one point the proposition it professes to prove, W. maintained that this assumption was quite unnecessary, and that the proof "rests entirely on the definition". I have an additional note on the facing page of my book which reads as follows:

"*Use of symbols and definition*

$(1) + 1 = 2 \quad 2 + 2 = ((1) + 1) + ((1) + 1)$

$((1) + 1) + 1 = 3 \quad$ By definition $a + (b + 1) = (a + b) + 1$

$(((1) + 1) + 1) + 1 = 4 \quad$ so $((1) + 1) + ((1) + 1) =$

$(((1) + 1) + 1) + 1 = 4$

The definition is really an infinite law of producing definitions." This may suggest that W. considered the initial definition enough to establish the rule for any given number; cf. Moore p. 7.

On the spirals cf. *Ph. R.* pp. 196–197.

Skolem's proof to which W. is referring is given in a footnote on p 194–5 of *Ph. R.*

W.'s engagement diary records two further lectures this term; my absence is explained by the fact that this was the term of my final Tripos examination—H.D.P.L.

SERIES B: 1930–31
Michaelmas Term 1930

Lecture B I

1. The nimbus of philosophy has been lost. For we now have a method of doing philosophy, and can speak of *skilful* philosophers. Compare the difference between alchemy and chemistry; chemistry has a method and we can speak of skilful chemists. But once a method has been found the opportunities for the expression of personality are correspondingly restricted. The tendency of our age is to restrict such opportunities; this is characteristic of an age of declining culture or without culture. A great man need be no less great in such periods, but philosophy is now being reduced to a matter of skill and the philosopher's nimbus is disappearing.

What is philosophy? An enquiry into the essence of the world? We want a final answer, or some description of the world, whether verifiable or not. We certainly can give a description of the world, including psychical states, and discover laws governing it. But we would still have left out much; e.g. we would have left out mathematics.

What we are in fact doing is to tidy up our notions, to make clear what *can* be said about the world. We are in a muddle about what can be said, and are trying to clear up that muddle.

This activity of clearing up is philosophy. We will therefore follow this instinct to clarify, and leave aside our initial question, What is philosophy? We start with a vague mental uneasiness, like that of a child asking "Why?". The child's question is not that of a mature person; it expresses puzzlement rather than a request for precise information. So philosophers ask "Why?" and "What?" without knowing clearly what their questions are. They are expressing a feeling of mental uneasiness.

The voice of instinct is always right *in some way*, but has not yet learnt to express itself exactly.

2. What is a proposition?

Propositions are the basic elements of our description of the world.[1] What are they? "Sentences in a book between two full stops"; or "Expressions that may be true or false". But "true" and "false" are superfluous. Thus I can answer the proposition "It is raining" by saying "True". But I might also say "It *is* raining". The repeated affirmation can stand for "true".

Any affirmation can be negated: if it has sense to say p it also has sense to say $\sim$ p. If you say "The electric lights are burning" when they are not, what you say is wrong (false) but it has meaning. Whereas if you say "'Twas brillig and the slithy toves" it has no meaning.

A proposition therefore is any expression which can be significantly negated.

3. But what is it to have sense and meaning? And what is negation? A proposition has sense, it may perhaps be said, if

---

[1]This sentence occurs in my notes only, where for "basic elements" I have the Greek στοιχεια which Wittgenstein certainly did not use, but which I must have throught expressed his meaning. Compare the proposition as the "unit" of language, pp. 42, 45, 57 below.

the words in it have meaning. But how do words come to have meaning?

(i) By definition. e.g. orange = yellowish red. But here we must how what "yellowish" and "red" *mean.*

(ii) By making people understand them, by causing certain processes to take place in people, for example by a drug.

(iii) By ostensive definition. But here all we are doing is to add to the symbolism. The ostensive definition does not get us away from symbolism. It is not final and can be misunderstood. All we can do in an ostensive definition is to replace one set of symbols by another. The result is a proposition, which can be true or false. Explanation of the meaning of symbols is itself given in symbols. What is essential in the explanation of a symbol to the understanding of the symbol sticks to it. The way we actually learn its meaning drops out of our future understanding of the symbol—the drug drops out in (ii) above.

4. Take the colour-word *green* as an example. The history of how we came to know what it means is irrelevant; what remains is our understanding. If we were to explain, without ostensive definition, to someone who knew nothing about colour, that orange means yellowish red, he would have to remember the phrase in order to understand the meaning in future. The word "green" may produce in us by association a memory-image of green. But the memory-image is not the meaning of green. The memory-image is as remote from the colour green as the word is. It is still a symbol, it does not bring us into contact with reality. To get to reality we must be able to compare the memory-image with reality, with an actual green patch. The possibility of comparison must be there even though there were no actual green things existing.

5. It may be said, A proposition is an expression of thought.

The expression need not be in words. We can substitute a plan for words. And a thought may be a wish or an order. Truth and falsehood then consist in obedience or disobedience to orders. Thinking means operating with plans. A thought is not the same thing as a plan because a thought needs no interpretation and a plan does. The plan (without its interpretation) corresponds to the particular sentence, the interpreted plan to the proposition. Proposition and judgment are the same thing, except that the proposition is the "type" of which the judgment (made in a particular place by a particular person at a particular time) is the token: cf. a number of copies of a plan or of a photograph. How do we know that someone had understood a plan or order? He can only show his understanding by translating it into other symbols. He may understand without obeying. But if he obeys he is again translating—i.e. by co-ordinating his action with the symbols. So understanding is really translation, whether into other symbols or into action.

We cannot get the interpretation into the plan; the rules for interpreting a plan are not part of the plan.

6.   In science you can compare what you are doing with, say, building a house. You must first lay a firm foundation; once it is laid it must not again be touched or moved. In philosophy we are not laying foundations but tidying up a room, in the process of which we have to touch everything a dozen times.

The only way to do philosophy is to do everything twice.

Lecture B II

1.   Hope, fear and doubt are forms of thought. Suppose I am expecting somebody to come to tea and he is an hour late. Is it one thought of an hour long I have about his being late,

or a sequence of thoughts? Does thought take a shorter time than its expression? (Words are clumsy instruments). Is thought something instantaneous starting before its expression, or is it a continuous state? Thought is a symbolic process and it lasts as long as its expression.

Consider the example of digestion. We can consider it

a) as a process characteristic of human be'ngs,

b) as a chemical process, quite apart from whether it happens in the stomach.

We take a similar view of thought. But any physiological process involved in thought is of no interest to us. Thought is a symbolic process, and thinking is interpreting a plan. It does not matter where this takes place, whether on paper or on a blackboard. It may involve images and these we think of as being "in the mind". This simile of "inside" or "outside" the mind is pernicious. It is derived from "in the head" when we think of ourselves as looking out from our heads and of thinking as something going on "in our head". But we then forget the picture and go on using language derived from it. Similarly, man's spirit was pictured as his breath, then the picture was forgotten but the language derived from it retained. We can only safely use such language if we consciously remember the picture when we use it.

Thought is a symbolic process. It does not matter a damn where it takes place, provided the symbolic process happens.

2.   Would it be possible to communicate more directly by a process of "thought reading"? What would we mean by "reading thought"? Language is not an indirect method of communication, to be contrasted with "direct" thought-reading. Thought-reading could only take place through the interpretation of symbols and so would be on the same level as language. It would not get rid of the symbolic process. The idea of reading a thought more directly is derived from the

idea that thought is a hidden process which it is the aim of the philosopher to penetrate. But there is no more direct way of reading thought than through language. Thought is not something hidden; it lies open to us. What we find out in philosophy is trivial; it does not teach us new facts, only science does that. But the proper synopsis of these trivialities is enormously difficult, and has immense importance. Philosophy is in fact the synopsis of trivialities.

3.   A plan is capable of different explanations. What does it mean to understand a plan? It means to be able to carry it out, to be able to obey it. What does I *can* carry it out mean? If I say I can lift a weight the only way to find out whether I can is to try to lift it. The "can" here is a hypothesis. But there is another sense of can in which there can be no doubt when I say that I can obey an order or that I can do something, that I can; even if subsequently I find I am mistaken (if after all I *cannot* recite the poem as I said I *could*).

4.   Distinguish (i) sign and (ii) symbol.

(i) the sign is the written scratch or the noise. We give the scratch or the noise—the word—meaning, with which it is used in the proposition which has sense.

(ii) everything which is necessary for the sign to become a symbol is part of the symbol, all the conditions necessary to give it sense or meaning are part of the symbol. These conditions are internal to the symbol and do not connect it with anything else. Explanation completes the symbol but does not (so to speak) go outside it. A sign can be nonsensical but a symbol cannot.

If we hear the words "I am tired" without seeing the speaker, they mean less than if we saw his lips moving and heard him say the same phrase. Thus the movement of the lips is part of the symbol, and the words "I am tired" written on the blackboard are an incomplete symbol. Anything that makes the sign significant is part of the symbol.

In any theory of meaning by association, which says that meaning consists in recalling something into the mind, the images, etc. recalled are themselves part of the symbolism, the association is part of the symbolism and works inside it. The words are not just reminders.

When we explain the meaning of a sign by pointing, we complete the symbol. We give further conditions necessary for understanding the symbol, which belong to the symbol. Explanations give us something which completes the symbol but does not supersede it, and this stays with us.

For a symbol to have meaning it is not necessary that the actual occasion of its explanation should be remembered. In fact it is possible to remember the occasion but to lose the meaning. One may remember that the meaning of "orange" was explained to one but not remember what colour it was. The image has got lost and we cannot understand what the word means. Similarly, I may have met Smith and be able to recognise him subsequently. But I may only remember him as the man I met in X's room: i.e. I may remember the occasion on which the meaning of the sign Smith was explained but not be able to recognise Smith, that is, not know the meaning of the sign.

The criterion of an explanation is whether the symbol explained is used properly in the future. And everything necessary to give the symbol meaning or sense is part of the symbol (cf. 4 (ii) above). The place of a symbol in language is shown by the way in which it is used.

5. What happens if one replaces one symbolism by another, and if all the signs are changed? How do we find the correspondence? If sign a has the same meaning in the new

symbolism as β in the old, then a has replaced β. But how do we know that a stands in the same place in the new symbolism as β in the old? The meaning of a word is its place in the symbolism, and its place will be shown by the way in which it is used in the new symbolism. One could substitute abracadabra for "green", leaving language otherwise unchanged, and discover its meaning by finding the way in which it was used. And one can often guess what the meaning of an unknown word is if it occurs in a proposition in which all the other words are known. The structure of symbolism must be such that the symbols in it cannot be substituted for each other without altering the geometry of the symbolism (cf. the colour octahedron; though this is not entirely unambiguous, as it can revolve on its W-B axis: see fig. 4). As its meaning is part of any symbol, every symbol must have its place unambiguously assigned. Symbols with different meanings must occupy asymmetrical positions.

There is no such thing as an insignificant symbol. By giving meaning to the sign I do not transcend the symbol but complete it.

Lecture BIII

1.   A French politician once said that French was the most perfect language because in French sentences the words followed exactly the sequence of the thought. The fallacy here is to think that there are, as it were, two series, a thought series (ideas, images) and a word series, with some relation between them. A further fallacy is to suppose that "thinking a proposition" means thinking its terms in a certain order. A proposition is a mechanism, not a heap or conglomeration of parts. The parts must be connected in a certain way, as they are in a motor-car, which is not just a box full of parts. We

think by means of the sign: to think of a thing is to think a proposition in which it occurs. The sign is not the cause of our thinking, and what causes us to think is not part of our thought, whereas the words *are* part of it.

The symbol is everything which is essential to the significance of the sign: compare the definition in Lecture B II 4 above. When we explain the meaning of a sign we are describing the symbol, not transcending it, the meaning is part of the symbol. If we explain the sign by ostensive definition, we are making clear the significance of the symbol.

2. We ask, What does negation mean? How can the word "not" express or explain negation? The word "not" is a kind of signal—it says "negate". But the signal needs an explanation. There must be some arrangement about its use. The red light is not by itself an order for the engine-driver to stop; it must have been explained to him in language. If the word "not" says, "negate this", then its use in this way can either be explained or not. If it can be explained, the explanation will be in words, and explanation ends when it has reached the greatest degree of explicitness. The meaning of "not" can only be expressed in rules applying to its usage. A proposition cannot be significant except in a system of propositions. If a man understands the order expressed in AB in fig. 7 he must also understand the order expressed in AC.

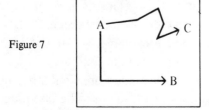

Figure 7

3.   But does not this make symbolism self-contained? Do we not rather say that the essence of symbolism is to point beyond itself? It seems to us as if the proposition was not simply itself, but pointed beyond itself and contained a kind of shadow of its fulfilment, which is neither the proposition nor the fulfilment but something between them (e.g. my expectation that Mr. Smith will come into the room seems to *foreshadow* his so doing). If you give an order it will not necessarily be obeyed, but does it not outline a shadowy fulfilment? For example, I tell someone to go to Matthew's to buy an orange. It seems that the sense of the proposition/ order, as opposed to the sentence expressing it, was a shadow standing between the proposition/order and its fulfilment.

What can this shadow be? If the order "Go to Matthew's and buy an orange" contains a shadowy fulfilment, the shadow is not the fulfilment, but something 'similar' to it in certain ways, it might be said. We look for this shadow particularly in an expectation. If I expect Mr. Smith to come in, my expectation contains something similar to Mr. Smith; if I expect a red patch my expectation cannot contain the red patch itself but only something similar to it. But to expect something "similar" is already to expect something different. The "similar" element, the shadow, in the expectation, is different from the fulfilment; you are no *closer* to the fulfilment.

Indeed, the expression "similar to" is already used up in language. For we can say, "This is similar to what I expected" or "This is exactly what I expected". But if it is exactly what I expected how can it be similar to it? The supposed shadow cannot be similar until there is something for it to be similar to; the use of the word similar presupposes that the two things are already there to be compared. Our symbols can never contain their own rule of projection or interpretation, and being similar is to be a projection of. The explanation by similarity won't do because one can't explain the similarity until both things being compared are there. An internal relation cannot be there unless both its terms are.

Suppose you ask someone "Do you know the alphabet?" He answers, "Yes". You ask, "Are you sure?" He runs over the alphabet in his mind and says, Yes, he is sure, though he has not repeated it out loud. This running over in his mind is the "shadow" of which we have spoken. But the shadow and the saying out loud are different. How do you know that what he says aloud is similar to what went on in his mind, what he imagined? The similarity only appears when he speaks. Was it there before he spoke? Of course not. All we can say is that he has projected the imagined alphabet into the "real" or spoken alphabet.

4.   In this context recognition is not the right word to use. The essence of recognition is that there should be some other test than the simple recognition; and if memory is the only test you cannot therefore talk of recognition in the true sense.

"Is this the same as you imagined?" "Yes; I recognise it". But how do you know you recognise it? We have here a question of the same kind as How do you know that another man sees the same as you do? There is no criterion for deciding and where there is no criterion we can't tell. Where there is a criterion, some answer is possible.

The same point is contained in the question, What is the difference between a portrait and a picture? What makes this picture a portrait of Mr. Smith? Is it similarity to Mr. Smith? No. The criterion of a portrait is not similarity because there are bad portraits and good. In a portrait we pre-suppose a particular kind of similarity; what makes a portrait a portrait is the intention. How can you explain the intention? Whatever one could add to the portrait could never make the intention clear, because some further interpretations could still be asked for. The rule of projection is expressed in projecting, the intention in intending. The internal relation is only there if both things related are there. You cannot anticipate the result which you project.

Lecture B IV

1.   Any explanation of a symbol can do nothing but add to the symbol. Yet our notion of a symbol is that it points outside itself, to a shadowy being between symbol and fact, the shadow of its fulfilment, which is meant to mediate between symbol and fact. What kind of thing must this shadow be? "It is similar to the fulfilment." This explanation we saw would not do. To interpolate a shadow or the like between symbol or proposition and fact will not do unless we have another shadow to mediate between that shadow and the fact; we have then an infinite regress, we are brought no nearer. As similarity does not explain the shadow, it does not explain the relation to fact either.

Part of the flaw here is that we confuse word and proposition. This confusion runs through all philosophy. If we say "Alpha, Beta", this does not pre-suppose the existence of the whole alphabet. Similarly, when we say the word "toothache" it does not stand for anyone having toothache, unless it occurs in a proposition.

2.   To return to "shadow" and "similarity", let us ask "What must an expectation be like to be an expectation of what is expressed in the proposition p?" The answer would be a description of the fact that would fulfil the expectation. But this has already been given in the proposition p. Can anything relevant be added to the content of the expectation already expressed in p? (We are not of course concerned

with what has caused the expectation.) No. Nothing can be added. We can of course analyse p: but that analysis will say the same as p. The only description of the content of the expectation is its expression in p. Anything that can add to or modify the content of the expectation must modify its expression.

What you now wish or expect is not a matter of future experience; for your wish or expectation may never be fulfilled. If I wish for an apple and find that a banana was what I really wanted, was my wish wrong? What kind of "similarity" are we looking for in the expectation? If one expects a green patch on the blackboard, then there must be a "green" constituent in the expectation, and the similarity of expectation to fulfilment is shown by the fact that both are expressed in the same words in language, and not by any further proposition.

3.   Instead of saying that the proposition must be similar to the fact fulfilling it, we might have asked how the fact *fits* the expectation and vice-versa. Take as an illustration a cylinder fitting into a cylindrical mould. Then the expression of their fitting is that the description of the inside and outside surfaces of mould and cylinder should be the same; the two surfaces have in common the equation which gives their shape. Expectation and fact meet, and what we are interested in is what they have in common; everything else is unessential.

Another illustration—someone sewing, either by hand or with a machine. The question then is what must the movement of hand (machine) be to produce stitches of a particular kind. (And what must our mental processes be in order that we can do what we do?) We can, of course, explain the mechanism of the muscles (or of the machine)—a physical explanation. But these don't interest us if we want to give the logical explanation. For that we can do without the fingers, the point of the needle, even the thread and its properties; nor if we are using a machine are we interested in its mechanism. What we are interested in is the geometry of the process of sewing. To produce such and such a stitch, the process must be so and so; the explanation of what the process of sewing must be like is contained in the shape of the thread in the linen. What is essential is what stitch and process (expectation and fact) have in common.

4. Language can express one method of projection as opposed to another. It cannot express what cannot be otherwise. We never arrive at fundamental propositions in the course of our investigations; we get to the boundary of language which stops us from asking further questions. We don't get to the bottom of things, but reach a point where we can go no further, where we cannot ask further questions. We do not resign ourselves to this, though we should see that we were wrong to ask the question. What is essential to the world cannot be *said about* the world; for then it could be otherwise, as any proposition can be negated. Our difficulty is that our intellectual discomfort is not removed until we have a synopsis of all the various trivialities. If one item necessary for the synopsis is lacking, we still feel that something is wrong.

Lecture B V

1. If we explain an event in physics, we explain it by describing another event. Thought is an event, and in psychology we can give a similar explanation of thoughts by describing other thoughts. If we say that in philosophy we don't want an *explanation* of thought but an *analysis*, this is

also misleading. When we analyse in science we describe some further event. In chemistry we analyse water and find that its chemical composition is H_2O; we find out something new about it. Analysing here means finding something new. But this is not what we mean by analysis in philosophy. In philosophy we know already all that we want to know; philosophical analysis does not give us any new facts. It is not the results of science which are of interest to philosophy but its methods. Philosophical analysis does not tell us anything new about thought (and if it did it would not interest us).

2.   Whatever necessary conditions I lay down for the fulfilment of an expectation must be added into the expression of the expectation, and the expression is the only thing that interests us as philosophers.

We tend to say we expect *something*; but our expectation would be correctly expressed in a *proposition*. We get confused and substitute a name—I expect Mr. Smith—and so get the idea that what we expect is a *thing*. Any expression which pervades our language like this causes deep-rooted and pernicious confusion. We confuse a name or word or meaning of a word ("red") with the proposition in which the word occurs ("This is red"); the word alone implies nothing about the world, the proposition says something about the world and can be negated.

3.   What expression and fulfilment have in common is *shown* by the use of the same expression to describe both what we expect and its fulfilment (the word "red" occurs in both).

But the phrase "in common" is itself confusing. "Have red in common" (for example) ordinarily means "Both are red"; which is not what we mean. This common element in expectation and fulfilment cannot be described or expressed in any proposition.

We might distinguish between having a property in common, and having a constituent in common, and say that we mean the latter. In that case the two uses must be entirely different, and the second must be a grammatical rule about symbols. For the constituent in common cannot simply be the word, the noise or scratch; it isn't just the noise "red" or the shape of the written word that matters. If we say that expectation and fulfilment have a constituent in common we are making a grammatical statement.

4.   What justifies us in using any particular word? Suppose I say "This gown is black". The word "black" is arbitrary in one sense; another sound or scratch would serve. And the correlation of the word "gown" to a particular object is in itself arbitrary and has no consequences. But if a proposition is to have sense we must commit ourselves to the use of the words in it. It is not a matter of association; that would not make language work at all. What is essential is that in using the word I commit myself to a rule of use. A word only has meaning in a grammatical system, and what characterises it is the way in which it is used. We can understand a plan only in relation to the system in which it occurs; cf. maps of the London Tube in which the stations are shown in the right order but as if they lay on a straight line.

5.   The method of projection must be contained in the process of projecting; the process of representation reaches up to what it represents by means of a rule of projection. If I copy anything the slips in my copy will be compensated for by my anger, regret, etc. at them. The total result—i.e. the copy *plus the intention*—is the equivalent of the original.

The actual result—the mere visible copy—does not represent the whole process of copying; we must include the intention. The *process* contains the rule, the *result* is not enough to describe the process.

We understand a symbol (the copy or the Tube map) as part of a system, and the system is described by its grammar (not by a further proposition).

Lecture B VI

1. What is "in common" between thought and reality must already be expressed in the expression of the thought. You cannot express it in a further proposition, and it is misleading to try. The "harmony" between thought and reality which philosophers speak of as "fundamental" is something we can't talk about, and so is not in the ordinary sense a harmony at all, since we cannot describe it. What makes it possible for us to judge rightly about the world also makes it possible for us to judge wrongly.

2. How can having a word in common be the expression of anything? (By word we do not mean the mere sound or scratch, but everything that makes a word a symbol.) A symbol cannot by itself be a symbol; what makes it a symbol is belonging to a system of symbols. A proposition is not a proposition unless it occurs within a grammatical system. If I use a symbol I must be committing myself; it is not just an arbitrary correlation of sounds and facts. If I say this is green, I must say that other green things are green too. I am committed to a future usage.

3. If I copy I must be following a rule of copying; some method of projection must be presupposed. (Again I am committing myself.) If there is any criterion for deciding

whether the copy is correct or incorrect, then there must be a rule of copying presupposed; otherwise we cannot talk of a copy at all. If [⟨ is to be a copy of [⟨ there must be a rule of projection to connect them. The rule *guides* me when I make the copy, and without it there can be no connection between copy and original—the "copy" could be anything.

4.   In a musical score the notes are clearly a picture of the notes to be played; they are a picture of the movement of my hands on the key-board. What of the sharp and flat signs? They are signals in the strict sense. Language does not consist of signals. A signal has to be explained, and the explanation must give you something which supplements the signal. The historical fact of the explanation is of no importance; what matters is what is given in the explanation. Even if we forget the explanation the signal may make us stop and ask what it means; we may remember we have seen it before but not what the explanation was.

If we are to understand it in future, the explanation of the signal must be sufficient; and if there are features of the signal that can't be explained they are irrelevant. But the explanation must be in language, that is, in symbols. That is all we have for the purpose (anything else would be magic).

5.   The notes and the sharp and flat signs in the musical score do not signify in the same way. The signs, being signals, must have been explained. The way in which we explain them is the way in which we explain colours. We need in addition to the word "green" something else. We can forget the explanation given us, and the word, the noise, "green" alone won't help you to find a pot of green paint. The word "green" (like the signal) must be connected by an explanation to a symbol in another language, e.g. the totally

different language of memory-images. But we are still using language (even if not *words*).

6.   Wherever there is misunderstanding we must appeal to an explanation. But the proposition in which the explanation is expressed is not different in kind from the explanation, even if it cannot be fully written or said. The explanation only adds to the symbol, which must then be enough in itself. If you understand a proposition you have nothing to help you but the proposition. We don't need to transcend it, but are guided by our understanding of it. Explanation of a proposition is always of the kind of a definition which replaces one set of symbols by another.

Lecture B VII

1.   That propositions have a word in common cannot express anything in so far as the word is arbitrary. As far as the word is arbitrary it can express nothing; what would be the use of the mere correlation of noises and facts? The correlation only has significance if we thereby commit ourselves to use the noises in the same way again. The correlation must have consequences, and language be to that extent rigid.

2.   It must be possible to be led by language. In what way? As we are led by the crotchets and quavers of a musical score. One is led by the position of the crotchet; "If the crotchets were elsewhere I would play differently".

      How do you know what you would do if the crotchet was differently placed?
(1) Not by experience. This is not something verifiable by subsequent experience. I must be able to know it *now*.
(2) Nor is it the same as to be mechanically led, as for

example by a pianola. The pianola may go wrong and we can't be sure it will not go wrong. And there is nothing in the machine itself that can be right or wrong. The machine goes as it goes.

But we are accustomed to look on a machine as the expression of a rule of movement. The machine per se never commits itself. We look on the machine as a symbol of a general rule; we see the *intention* behind it, the way it *ought* to work.

So to be "led by" crotchets means to follow a general rule. This rule is not contained in the result of playing, nor in the result plus the score (for the score might fit *any* playing by *some* rule). The general rule is contained only in the *intention* to play the score. A description of the act of playing will in no way contain a description of the score; but a description of the intention will contain a description of the score ("He tried to play according to this score"). We *see* the rule in the relation between playing and score.

3. What is meant by "contain"?

The series 1,4,9,16 . . . can be interpreted according to a general rule, without the general rule being written down (e.g. in the form x/x^2). The general rule is contained in all special cases, and so cannot be isolated. The general member of the series (x/x^2) does not isolate the general rule, for we could go on to ask, "How is the general member applied?" The general member can only be understood by seeing how it is used. If I see a general rule I see it in a special case; and when I interpret the general member I am doing the same thing as I do when I interpret any other special case. The special case contains the general rule but not the explanation of the rule. The rule is what makes the symbolism not arbitrary. We cannot verify it by experience, but it is contained in our intention. The general rule is the standard in terms of which we judge what we are doing (whether we have played the score rightly or wrongly; and if the player does not actually *read* the score, that is irrelevant).

4.   Things are similar only with reference to a rule of projection. This is often obscured by the fact that we generally use only one rule, that of the portrait which must be similar to its original. But traffic movements are similar to the red, amber, green of the traffic lights. The choice of colours is arbitrary, but once we adopt it we are committed. A rule of interpretation limits the possibility of significance in a symbolism, it provides for certain possibilities and not for others. These others are then meaningless until we provide a new rule (we might add a sign to our musical notation which meant "Play twice as loud"). The rules of a symbolism give it a certain degree of freedom which is expressed in the rules of its grammar, which tells us which combinations are allowed and which are not allowed. Explanation increases the multiplicity of the system by distinguishing between possible interpretations. Understanding a symbolism is correlated to an explanation, and explanation removes *mis*understanding. When the right multiplicity has been given, no further explanation is wanted or possible.

Lent Term 1931

Lecture B VIII

Recapitulation

1. The answers which philosophy gives to our questions must be fundamental to everyday life and to science. They must be independent of the experimental findings of science. Science builds a house with bricks which, once laid, are not touched again. Philosophy tidies a room and so has to handle things many times. The essence of its procedure is that it starts with a mess; we don't mind being hazy so long as the haze gradually clears.

2. We started with propositions (sentences). Propositions seem to be the units of our description of the world, and when we asked what is a proposition we were expressing a vague mental discomfort, like the child's "Why?"

What is a proposition? A sentence between two full stops? An expression of thought? A description of fact? A statement of what is the case? A statement that can be true or false?

It is not an *expression* in the sense that e.g. crying is an expression of pain. Nor is it a series of hallucinations: e.g. "It is fine weather" is not a hallucination of "fine" followed by one of "weather." Nor is it momentary or amorphous. And we are not interested in *thought* from a psychological point of view, in its conditions, causes and effects; we are interested in thought as a symbolic process. Thought is an activity which we perform by the expression of it, and lasts as long as its expression.

The sentence is not just a series of words which bring before the mind a series of pictures by association. This is an idea derived from proper names, which are thought to "mean" the person to whom they refer. The question "What is the meaning of a word?" could be answered by saying "*This* is the meaning", if that were the case. But when we say that the word "and" has meaning what we mean is that it works in a sentence and is not just a flourish. This may be called the intransitive sense of meaning, when there is no corresponding existing thing. But the words "red" and "green" do not require the existence of red or green things in order to have meaning. My idea or image of red is not red, and "I expect a red patch" has meaning even though I have never seen a red thing and everything red in the universe has been destroyed.

If in some cases the meaning of a word is given by ostensive definition, i.e. if there is something I point to, then I have action instead of words, but I still have a sentence, and I cannot ask for further explanation. I cannot explain the ostensive definition, and I have not replaced the proposition (e.g. I shall eat an orange in five minutes) by the fact; I have replaced one symbol by another but not by the symbolised. In fact an ostensive definition works in the same way as any other symbol. A symbol is a sign together with all the conditions necessary to give it significance. "Understanding" means getting hold of the symbol, not the fact; and understanding is what is conveyed by an explanation (not by a drug or external agency). Explanation adds to the symbol, gives us more to get hold of. The symbol is in some sense self-contained; you grasp it as a whole. It does not point to something outside itself, it does not anticipate something else in a shadowy way. Nor does understanding a symbol imply a knowledge of whether it is true or false; for that the fact is required also. But it is generally supposed that the thought (belief, wish) *foreshadows* its verification; and this shadow we call the meaning of the proposition, and it seems to mediate between the proposition or symbol and the fact. But this assumption of an intermediate link does not help us, as we should need another link between the shadow and reality, and so on indefinitely.

The shadow is supposed to be *similar* to the fulfilment. But there is no absolute similarity. Similarity varies according to the mode of projection; but we tend to assume, when we use the word, one particular mode of projection, the mode we assume in looking at a photograph. On this view we could hardly call my image in a concave mirror "similar" to me. If my expectation is a shadow of the fact expected, how can I expect Mr. Smith to enter the room if he does not? And if it is replied that my expectation is similar to Mr. Smith, the answer is that what I expect is Mr. Smith, and just Mr. Smith. What is true is that the expression of my expectation contains a description of the fact which satisfies it when it comes.

3.   To understand a thought means to be able to translate it according to a general rule. For example, playing a piano from a score. But the score does not *cause* us to play as we do; if it did there would be no right and wrong way of playing.

If a, b, c and d are respectively defined as $\longrightarrow$, $\uparrow$, $\longleftarrow$, $\downarrow$; then a a d a b c is $\longrightarrow$ $\longrightarrow$ $\downarrow$ $\longleftarrow$ $\uparrow$ $\downarrow$ $\longrightarrow$

At first sight we would say there was no similarity between the letters and arrows. To the rules of translating there correspond the rules of grammar, and for these there is no justification. The language in which we might try to justify the rules of grammar of our language would have to have a grammar itself. No description of the world can justify the rules of grammar.

Lecture B IX

1. The proposition is the unit of what can be said. A proposition is a description of fact, of what is the case, and is either true or false.

2. To say that a word has meaning does not imply that is *stands for* or *represents* a thing. Such words as "and", "not", "or", etc. obviously do not stand for anything. And the ostensive definition which gives the meaning of a proper name is itself a symbol and cannot be replaced by anything else. But, it may be asked, even if words do not stand for or represent things, cannot thought do so? Is not this the peculiar property of mental phenomena? Is there not representation "in the mind"? This suggestion is a pernicious mistake. It separates thought into two parts, organic (essential) and inorganic (non-essential). But no part of thought is more organic than another. There is no mental process which cannot be symbolised; and if there were such a process which could not take place on the blackboard, it would not help. For I could still ask for a description of this process, and the description would be in symbols which would have a relation to reality. We are only interested in what can be symbolised.

3. Thought is autonomous. If you say "Townsend is sitting on the bench", Townsend, sitting and bench cannot be *in* your mind. But it might be suggested that each of them sends a representative to your mind. And there is some truth in the suggestion. But what guarantee could we have that they represent *anything*? What is given in my thought is there and is essential to it. Everything else (e.g. anything *represented* in my thought) is irrelevant.

4. Thought is therefore autonomous, complete in itself;

and anything not given in my thought cannot be essential to it. Thought does not point outside itself. We think it does because of the way in which we use symbols. We compare the symbol with something else, or we translate it into a description in other terms or an action within our control. A description must be entirely determined by the fact described plus the grammar and vocabulary of the language used. I can *choose* the language which I use, but my description is then determined by the grammar and vocabulary of the language chosen.

In so far as we are concerned to ensure understanding, we either add something to the sign to complete the symbol or give a translation of some kind (e.g. into imagery). The translation is made according to a general rule by which we derive the result from the first symbol.

Anything that can be decided must be capable of being symbolised.

5. Behaviourism must be able to distinguish between real toothache and simulated toothache, between a man who is pretending to have toothache and a man who really has it. If I see you reading I can only say that you have certain symbols before you and do something. But I must be able to distinguish between reading and not reading. And the text may cause me to read correctly or incorrectly, so reading cannot be simply a matter of causation.

6. I cannot prescribe the use of symbolism as such; to do so would require another symbolism. I can prescribe the use of *a* symbol, but only by adding further symbols. *A* language can be taught by a correspondence course, language cannot. All I can do to clear up symbolism is to describe the symbolism. Grammar (rules and vocabulary) is the description of language, and it consists in giving the rules for the

combination of symbols, i.e. which combinations make sense and which don't, which are allowed and which are not allowed.

Can we give a description which will justify the rules of grammar? Can we say why we must use *these* rules? Our justification could only take the form of saying "As reality is so and so, the rules must be such and such". But this presupposes that I could say "If reality were otherwise, then the rules of grammar would be otherwise". But in order to describe a reality in which grammar was otherwise I would have to use the very combinations which grammar forbids. The rules of grammar distinguish sense and nonsense and if I use the forbidden combinations I talk nonsense.

If grammar says that you cannot say that a sound is red, it means not that it is false to say so but that it is nonsense—i.e. not language at all. Therefore I cannot say that sounds have properties which colours have not, because I should then have to be able to say significantly that colours have properties which they have not. To call a thing a colour is to say it obeys certain grammatical rules.

7. A proposition is a logical form and I cannot therefore define it as characterised by certain properties which distinguish it from something else which has not these properties; for I could then significantly deny that it had these properties. A proposition is simply characterised by the grammatical rules which apply to it. So to say that a proposition is what can be true or false is to say that a proposition is anything which grammar allows me to call true or false. Similarly, we can say that a proposition is anything that can be negated, any expression to which grammar allows me to apply the symbolism of negation. Negation in propositions and negation in mathematical equations have certain grammatical rules in common; but

there are other rules which propositions and equations do not have in common.

Lecture B X

1.   Grammar circumscribes language. A combination of words which does not make sense does not belong to language. Sense and nonsense have nothing in common. By nonsense we mean unmeaning scratches or sounds or combinations.

2.   What does it mean to use language according to grammatical rules? It does not mean that the rules of grammar run in our heads as we use language. There is no need to repeat the rule. But there must be rules, for language must be systematic. Compare games: if there are no rules there is no game, and chess, for example, is like a language in this sense. When we use language we choose words to fit the occasion.

3.   An explanation gives understanding as opposed to misunderstanding. It cannot teach you understanding as such, it cannot create understanding. Explanation makes further distinctions, i.e. it increases multiplicity. And when multiplicity is complete, then there is no further misunderstanding and no further explanation is required.

The whole of language cannot be misunderstood, because if it could be there would be no possibility of an explanation to make one understand it. Every word in language must have its place in language independently of what is arbitrary about the sign. Its place is determined by all the grammatical rules which apply to it, i.e. by all explanations of it. So the king in chess has a place in the game determined by the rules and quite independent of its actual shape. What determines

the place of a word in language is all the explanations of it
that can be given previous to its use; and the answer to the
question "Why did you chose this word?" will be a
grammatical rule governing its use.

4.　There may of course be symbols which of themselves
form a system. If my definition of colours is given by means
of coloured patches, then these form a system on their own in
a way which colour words do not. Again, we choose
onomatopoeic words because they too belong to a system
(rustle, boom, hiss), and the word itself is a picture.

5.　You cannot justify grammar. For such a justification
would have to be in the form of a description of the world
and such a description might be otherwise, and the
propositions expressing this different description would have
to be false. But grammar requires them to be senseless.
Grammar allows us to talk of a higher degree of sweetness,
but not of a higher degree of identity; it allows the one
combination but not the other, nor does it allow us to use
"sweet" instead of "great" or "small". Is grammar arbitrary? Yes, in the sense just mentioned, that it cannot be
justified. But it is not arbitrary in so far as it is not arbitrary
what rules of grammar I can make use of. Grammar
described by itself is arbitrary; what makes it not arbitrary is
its use. A word can be used in one sense in one grammatical
system, in another in another.

6.　A proposition is, then, a logical form, characterised by
certain grammatical rules. And what characterises propositions as commonly understood is that truth functions make
sense with them. A proposition is whatever can be true or
false; it states that 'so and so is the case' or 'is not the case'.
This common use includes hypotheses and mathematical

propositions as well as propositions in the strict sense. "There seems to be a man here" is a proposition. "There is a man here" is a hypothesis. When we distinguish propositions from hypotheses and mathematical propositions we mean that some grammatical rules apply to the one and not to the others.

7.   True and false are nothing but a part of the notion we call truth functions. They are indicated by such words as yes, no, and, or. p is true = p: p is false = ~p. Frege explained such truth functions (~, or, and etc.) by propositions containing the words true and false; and this explanation is itself a notation which can be substituted for them. (Every explanation can be substituted for what it explains.) It is not a statement of a relation between e.g. ~ p and p is false but a definition.

$$\left.\begin{array}{c|c} p & \\ \hline T & F \\ \hline F & T \end{array}\right\} = \sim p$$

But not "If 'p is T' is F, and 'p is F' is T, then ~p". It is a definition not an expression of an external relation.

Every such explanation is a definition; a logical explanation is quite different from a physical explanation.

Lecture B XI

1.   We justify the use of a word by giving a rule, just as we would justify a move in chess by giving a rule. This does not mean that we have the rule running in our head all the time. We use language without conscious awareness of its rules; but the rules justify the use. To make a move in chess

according to a rule is quite different from making it at random. It does not matter whether the account of making the move is behaviouristic or introspective; both must have the same multiplicity. But the man who makes the move in chess according to rule sees something different from the man who does not. Similarly, the man who understands a word sees more in it than the man who does not; so his mental process will differ, and perhaps his intonation.

2.   The process involved in understanding a proposition is the same as that involved in understanding a picture or a picture-story. We see a picture on a flat surface as three-dimensional. Similarly we can see fig. 8 in more than one way. It is not a matter of interpretation; we *see* something different. So also we experience something different when we hear and understand from what we experience when we hear and do not understand. You actually get different impressions in both cases, pictures and words.

Figure 8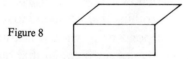

We are not interested in these processes themselves. For us explanation and understanding are the same, understanding being the correlate of explanation, which produces it (though *not* causally), and having the same multiplicity. In a definition the explanation is substituted for what is explained. It is a longer way of putting it. So we said above that the behaviouristic and the introspective account of a move in chess must both have the same multiplicity, and multiplicity is what we are interested in.

3.   A proposition is an expression to which the rules of the True and False game apply in a particular way. T and F can

only be correctly used as part of a notation for truth-functions, and indeed we could do away with them entirely if we define p as meaning p is true and ~p as p is false. But the important point about the T and F notation is that it can be substituted for Russell's truth-function notation. Russell explained ~p by saying that ~p is true if p is false and false if p is true. But this is not an explanation of negation, for it might apply to propositions other than negative. We don't want an explanation of the kind that Russell and Frege try to give. (Frege's explanation only works if words can be substituted for his symbols.) What we want is an "analysis" or definition. Negation is more complicated than the ~ sign.

To say that ~ or "not" is indefinable satisfies no one; nor does a definition (e.g. p/p = ~p: definition) satisfy either. To give a definition is simply to state one of the rules about it. And if you know the rules governing the use of ~p you are not absolved by saying you can't define it.

4.   It is sometimes asked, how can we negate a proposition? For if ~p is true there is nothing to correspond to it; there is only something to correspond to p when p is true. What corresponds to 'the door is not open' when it *is* open? But there is a false analogy here; p does not correspond to *something*. And what corresponds to ~p is p *not being the case*.

5.   Implication

Russell said that "p implies q" is true even though p is false. In the T—F notation

p	q	
T	T	T
T	F	F
F	T	T
F	F	T

This sounds paradoxical for two reasons:

(1) We confuse implication and inference. A true proposition cannot be inferred from a false as Russell's notation seems to suggest.

(2) The sign p⊃q has been translated If p then q. But in ordinary life we never use "If . . . then" in this sense. Nor do we use "If . . . then" for the expressions which Russell symbolises by (x)φx⊃ψx. We use it always in hypotheses. A proposition is verified or falsified in experience. But a hypothesis is not capable of definite verification in experience. "Propositions" about physical objects and most of the things we talk about in ordinary life are always really hypotheses. For example—All men are mortal. As Russell uses (x)φx⊃ψx it remains true even if there is nothing answering to the description ψx. So All men are mortal, remains true even if there are no objects answering to the dscription "men", i.e. if ψx is never true. But we certainly never mean this when we use such expressions. We do not mean that all men are mortal even if there are no men. In other words, by "If . . . then" we mean something quite different from implication in the sense in which Russell defines it.

Lecture B XII

1. If asked for an explanation of negation we can reply "But don't you know what it is?" We could not reply similarly if asked for an explanation in physics. We know how to use the word "not"; the trouble comes when we try to make the rules of its use explicit. Correct use does not imply the ability to make the rules explicit. Understanding "not" is like understanding a move in chess. We see something different in a move when we understand the rule; compare also seeing something different in fig. 8 in Lecture XI.

2.   When I say "This door is not open" and "The chair is not yellow" is it the same *not* in both sentences? How do we know that *not* and *nicht* mean the same? (1) We may know through a definition: *nicht* = *not*. (2) But we might have learned both languages and in the process somehow gathered that both words had the same meaning. Learning a language makes us see something different in the sentences of the language from what we would see if we had not learned it. The *process* of learning does not matter; it is history and history does not matter here. There is an act of laying down a rule; for example we might lay down the rule on going for a walk to toss for it whenever we had a choice of ways. This laying down of a rule is exactly analogous to learning language. The laying down of the rule is not contained in following the rule; the laying down is history.

So the rules applying to "not" describe the understanding of "not"; but the laying down of the rules does not enter into the understanding of "not".

3.   In logic nothing is hidden. We can't get any clearer about the use of "not" by discovering something new about it. What is difficult is to make the rules explicit. If I say, in terms of the diagram in Fig. 9, that if he understands path AC he must also understand path BC I am describing his understanding. BC need not enter his understanding of AC, but he must understand in a system. We see in a different way according to the system in which we see. The rules actually describe our understanding of a sign.

Figure 9

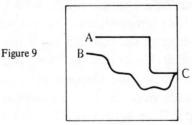

Compare—

System A	$\longrightarrow$	move in any direction at speed S
	$\longrightarrow$	move in any direction at speed 2S
System B	$\longrightarrow$	move in this direction at any speed

Then——$\longrightarrow$may be a symbol in either system.

4. How can not-p have meaning? For when not-p is true there is nothing for not-p to refer to. There is a typical puzzle due to a false analogy. As a result of this false analogy it has been suggested that we should substitute for a negation a disjunction of all the other possibilities. This will work in some cases: e.g. this is a primary colour but not red=this is green or blue or yellow. But even if we allow this (and it is quite an important concession) not-p will none the less have the same multiplicity as the disjunction. The excluded possibility p, plus the sign of exclusion ($\sim$) can stand for the disjunction and have its multiplicity

Lecture B XIII

1. "Do not run away, and this negative is such that a double negative = an affirmative". The second part of this sentence can be nothing else than a rule; and if the first part has sense the second part is already known. The rule says nothing about negation, but does say something about the use of "not" (or the sign $\sim$).

We cannot *describe* a cube or circle in geometry; we can define them. We can describe the shape of a patch or a body, but not of a cube or circle. Geometry describes a circle to the same extent as logic describes negation. Geometry gives the grammar of certain connexions, and the multiplicity of these connexions does correspond to what we mean by cube, circle, etc. So $\sim p = \sim\sim\sim p$: and so on. The rule which allows this must give the essence of negation: the rule characterises negation.

2. Inference. Treat this with the T and F notation. A tautology says nothing; if you add a tautology to a proposition you add nothing to what the proposition says. (In symbols—p.tautology = p; similarly p. contradiction = p.) A proposition gives reality a degree of freedom; it draws a line round the facts which agree with it, and distinguishes them from those which do not. In the T F notation, T F F F gives less freedom than T F F T. But the tautology gives all degrees of freedom and so says nothing.

p	q	
T	T	T
T	F	T
F	T	T
F	F	T

This gives all degrees of freedom and so makes no connection between p and q.

Inference is the transition from one proposition to another, a transition which we justify by saying that e.g. q follows from p. This relation is entirely determined when the two propositions are given. It is entirely different from other relations, in which the opposite case is always thinkable. The relation of following and similar relations are internal relations which hold when (roughly) it is unthinkable that they should not hold. Whether a proposition is true or false can only be decided by comparison with reality. So that p v q follows from p.q is not a proposition: it has no use. What justifies the inference is seeing the internal relation. No rule of inference is needed to justify the inference, since if it were I would need another rule to justify the rule and that would lead to an infinite regress. We must see the internal relation.

Lecture B XIV

1. An internal relation holds by virtue of the terms being what they are. Inference is justified by an internal relation which we see; the only justification of the transition is our looking at the two terms and seeing the internal relation between them.

2. Russell's notation does not make internal relations clear. It is not clear in his notation that p v q follows from p.q. In Sheffer's stroke function the internal relation is made clear. p|q .|. p|q follows from p|p .|. q|q. In the T F notation we can however say that if the Ts of one proposition include those of another then the second follows from the first. Is not this a rule? It is a rule of grammar dealing with symbols alone, it is a rule of a game. Its importance lies in its application; we use it in our language. When we talk about propositions following from each other we are talking of a game. Propositions do not follow from one another as such; they simply are what they are. We can only prepare language for its usage; we can only describe it as long as we do not yet regard it as language. The rules prepare for the game which may afterwards be used as a language. Only when the rules are fixed can I use the game as a language.

To a necessity in the world there corresponds an arbitrary rule in language.

There is no smaller unit in language than the proposition; it is the first unit that has sense and you cannot build it up from other units that already have sense. When you state the rules of grammar you are still building, and language is not complete. Only when the rules are complete is it a language in which there are propositions. The rules of grammar are arbitrary in the sense that the rules of a game are arbitrary. We can make them differently. But then it is a different game.

3.   In a calculus we always use signs, though the calculus is quite independent of the particular scratches or other signs we use. Words are not arbitrarily chosen to remind you of certain characteristics. A calculus is never in itself right or wrong, but it may be rightly or wrongly used.

4.   The $\therefore$ which we write in p $\therefore$ p v q is a sign of the same kind as $=$; it is about symbols. Russell expressed p $\therefore$ q in the form $\vdash p \supset \vdash q$. It is indeed essential to a propositional sign that its beginning and end should be marked; there is no sense in writing p v q unless you know that that is all. Russell's $\vdash$ serves as a full stop; that is its only justification. To call it an "assertion sign" is misleading and suggests some sort of psychological process. If we talk of "asserted propositions" all we mean is the whole proposition (sentence) between the full stops. To write $\vdash p \supset \vdash q$ is therefore absurd. A tautology can occur in another proposition as an "unasserted proposition".

Lecture B XV

1.   Grammatical rules are arbitrary, but their application is *not*. There cannot therefore be discussion about whether this set of rules or another are the correct rules for the word "not"; for unless the grammatical rules are given "not" has no meaning at all. When you change the grammatical rules you change the meaning of the word. You cannot describe negation in terms of negation because that presupposes that you already know what the meaning of negation is.

2.   The use of predicates in logic is always misleading, suggesting as it does different "kinds" of term etc. differentiated by predicates. For example "formally certified",

"internal relations". Description by means of predicates must be capable of being otherwise.

Instead, give a description of symbols, or rather of *signs*. What we describe is the signs. The sign plus the rules of grammar applying to it is all we need. We need nothing further to make the connection with reality. If we did we should need something to connect that with reality, which would lead to an infinite regress. And an infinite regress means that you cannot reach a final point and so get no further. We don't need further signs to connect our signs with reality; all we can do is to give their grammar as opposed to other signs.

3.   The grammatical rules applying to it determine the meaning of a word. Its meaning is not something else, some object to which it corresponds or does not correspond. The word carries its meaning with it; it has a grammatical body behind it, so to speak. Its meaning cannot be something else which may not be known. It does not carry its grammatical rules with it. They describe its usage subsequently.

Easter Term 1931

My notes for this term are far less systematic than those for previous terms. They are for the most part taken in pencil, and there is nothing to mark the break between lectures. There are no notes by J.E.K. covering the term. At this time (see Introduction p. 00) W. moved his lecture class from the Art School to his rooms in Whewell's Court in Trinity and the change in style in my notes marks that change, though I have no clear recollection of it.

Justifying a usage, statement, etc. is connected with being guided by a rule. Justification essentially involves a variable.

To follow an arrow involves the possibility of an explanation containing a variable. (This is a grammatical rule about the word follow).

Distinguish between a cause and a motive/reason as an explanation. (This is a grammatical distinction). It is nonsense to ask how you know what your motive/reason is.

To say that a proposition is a picture stresses a certain aspect of the grammar of the word proposition.

"Arbitrary" as we normally use it always has reference to some practical end: e.g. if I want to make an efficient boiler I *must* fulfil certain specifications, but it is quite arbitrary what colour I paint it.

Experience is not distinguished from what is not experience by predicates. It is a logical concept, not a concept like chair or table.

The puzzles about time are due to the analogy between time and motion. There is an analogy, but we press it too far; we are tempted by it to talk nonsense. We say time "flows", and then ask where to and where from, and so on (cf. also analogies from the cinema film; the pictures are all present in the film, yet when it is shown have a relation of past and future). These questions are illegitimate. We solve the puzzle by giving the correct grammatical rules. The importance of these puzzles and rules is the importance of language.

Language can only be important because of the use made of it. It has no sense to say that language is "important" or "necessary" to communicate our meaning. But it may be important for building bridges and doing similar things.

If we can talk of using a sign twice with the same meaning, this meaning must be laid down somehow. The problem is that of recognition (Frege). But here recognition must be autonomous, which in the ordinary sense it is not, as ordinarily we acknowledge other criteria—as, e.g. if I recognise Mr. Smith. But if I say "This is brown" my recognition is the only criterion, not one of several. Compare Russell on memory.

It seems that though words, or even a painted picture, might need further interpretation the mental image does not: what I imagine cannot be misunderstood. Further interpretation is not possible in so far as the description of the image coincides with the description of the thing imagined.

We think that black is more closely connected with the fact that there *is* a black patch than with the fact that there is *not*. But it is not.

"We can't have an image without knowing what it is an image *of*." But why? How do you know? All you have is the image. It is like black and the black patch.

You cannot describe a calculus without using it, you cannot describe language without giving its meaning. (This excludes of course mere sounds and scratches.) The place of a word in logical space fixed by grammar *is* its meaning. You cannot say that in order that a word should be used as it is it must have *these* rules. The meaning of a word is given if you describe language by all its rules. All explanations take place *inside* language. They would only transcend language if they made assertions of fact, which they do not. The meanings of the words are part of language.

Only events, facts are particulars.

The relation of expectation and fulfilment is precisely that of calculation and result. 2×2, result 4. An action is the last step in a calculation. The calculation enables us to say which result is right and which wrong, but says nothing about what we will in fact write down.

We confuse the meaning of a word and the bearer of a name. The name is in a sense a substitute for or representative of its bearer. But it is not a substitute for its meaning. The meaning of the words Professor Moore is not a certain human body, because we do not say that the meaning sits on the sofa, and the words occur in the proposition "Professor Moore does not exist". A substitute cannot be a substitute for nothing. If Moore is a substitute then it must be true that Moore exists (i.e. that such and such is the case). "There are no red things" can only make sense if the word "red" has meaning, and the word red can only be a substitute if certain facts are the case.

What this comes to is this—

Meaning is fixed *inside* language, by explanations. The phrase "meaning of" is misleading as it suggests "representative of" or "substitute for". Meaning is *not* the object pointed to in an ostensive definition (cf. the history of our learning drops out). Compare what was said above about the relation of thought and reality being that of calculation and conclusion. Thought anticipates its fulfilment in the same sense that a calculation anticipates its fulfilment.

Mathematics can be learned beforehand, and so can language. So philosophy gives rules once for all.

R.D.T.'s continuous notes of the lectures run out at the beginning of Lent Term '31 Lecture B X. But there are in addition a few more pages of more disconnected notes. These must have been taken during that term and the last of them corresponds closely to the last paragraph of Lecture B XV and has been incorporated in it. But there is no obvious place where the remainder could fit in, and they are reproduced below as an appendix to that term's lectures.

Doing philosophy may perhaps mean resignation of temperament but never of intellect.

The fascination of philosophy lies in paradox and mystery.

Argument. Its purpose is not to convince positively, or convert, but to remove erroneous notions and prejudices.

A proposition has sense if we know what state of affairs will agree with it and what state of affairs will not agree with it.

"A proposition is a picture of reality"—but only if we do not take "picture of" to mean "similar to" in the ordinary sense.

A sign is needed only when there are other signs which could take its place. A rule is stated only where we could have other rules.

Explanation is merely substituting one sign for another. The symbol is self-contained—it does not *mention* something but does *contain* it. Whatever it might *merely mention* (as though by representation) might not exist, so it cannot be what we mean. My *understanding* a symbol can't be the *result* of the symbol as cause—the cause is outside the effect but everything necessary for the meaning of the symbol must enter it.

Every explanation of a sign is a definition—you can't do more than define, i.e., make a substitution, which does not lead outside language. This is also true of explanations of proper names by ostensive definition.

We are apt to think that the meaning of words lies outside language (an image is known as an image *of* something else), but the meaning of a word must be entirely given, or determined, if you describe the language or its rules.

SERIES C

Academic Year 1931–32

As explained in the Introduction, it has not been possible to identify breaks between sessions in this series. The headings under which the notes have been grouped are given below. With the exception of the comments on Broad, the text follows the order of the notes as taken, with a few minor transfers which are enclosed by † †, and some grouping of consecutive sentences into paragraphs.

C I   Philosophy, Proposition and Meaning

Philosophy describes what it has sense to say and what it has not sense to say.

A proposition is a judgement about sense data, a reading of one's sense-data; for example "This is red". No further verification is needed; it is *a priori*. A hypothesis is an expression of the form "This man is ill", "The sun will rise tomorrow" or "This is a chair". It is confirmed or rejected, when its meaning is clear, by empirical science.

The meaning of a proposition is the mode of its verification; † two propositions cannot have the same verification. †

The meaning of a word lies entirely in its use, and is given in an explanation. Words by themselves do not speak to you; † a word only has meaning in a proposition. †

Ethical and aesthetic judgements are not propositions because they cannot be verified.

Meanings are learned by translation; in childhood from gestures into words. But one did not have to learn what the gestures meant; they are explained by words.

To find out what words mean, you must see what is meant by the explanation of their meaning.

C II   Symbols, Primary and Secondary

"This is grey". This can be either a) a sentence or proposition, or b) a rule or definition of the use of language.

In a pattern book a) the primary symbol is the pattern, b) the secondary symbol is the number of the pattern. The numbers of the patterns are really names, for factory hands do not quote the type of pattern but the number which differentiates one pattern from another. This is how we use language. In all language there is a bridge between the sign and its application. No one can make this for us; we have to bridge the gap ourselves. No explanation ever saves the jump, because any further explanation will itself need a jump. No reason compels us to learn language.

If numbers are used for the patterns, the patterns are primary. If we call patterns and numbers signs, then if we say we are making copies of them, we are using an entirely different sense of the word copying when we apply it to patterns and when we apply it to numbers. In the sense in which we copy the patterns we do not copy the numbers.

If you do copy, is it fixed *how* you copy? No; not at all. You might copy the pattern twice as large as the original, and so on. In Fig. 1 there is nothing to compel you to write x for a. There is no reason why you should not copy according to

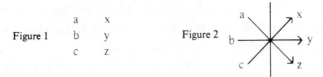

Figure 1

a   x
b   y
c   z

Figure 2

Fig. 2. The explanation needs a bridge again; it does not itself bridge the gulf. So in reading we don't make use of a table between the letter 'a' and the sound 'a'. We pass straight from the scratch to the sound. We look up nothing and it would not help us if we did; we should still have to make the jump. The sound is only associated with the letter 'a' when we have associated it.

What is significant is that you can give an order not only to 'take a piece of red chalk', but also to 'imagine a spot of red'. When 'imagining a red spot' you don't look up a pattern book and then proceed to imagine.

You can only explain the method of projection by projecting; and the projection is only made when it is made.

Every transition in language is as important and on the same level as a transition from language to action. Primary and secondary symbols are primary and secondary transitions.

C III   Visual and Physical Space

Visual and physical space are different in the sense that what you can say about the one you cannot say about the other.

Of visual space, it has no sense to say that anything looks farther away than the moon. But it has sense to say that something looks farther away than the moon in physical space. If the moon were seen to become smaller, we could not say that it was going farther away in visual space, but we could and should say so in physical space. The distinction is that between sense-datum and physical object.

Visual space and physical space are not distinguished as two species of the same genus, but are distinguished by the fact that there are some things which it has sense to say about both, and some it has sense to say of one and not of the other. So, (1) it has sense to say of both spaces that A is the same distance from B and C; (2) in physical space I can say "I see something further away than Sirius" or "I can see a star approaching or receding from us", but I cannot say this about visual space.

† If you try to define a penny as a class of sense-data, how can you describe the class and what is its relation to visual sensations? One always has an uncomfortable feeling that one is trying to do too much with too little material. †

There is no need of a theory to reconcile what we know about sense data and what we believe about physical objects, because part of what we mean by saying that a penny is round is that we see it as elliptical in such and such conditions.

It has no sense to say that an object at the edge of the visual field is perfectly clear, just as one cannot talk of greenish-red.

In visual space you cannot see a triangle as being equiangular and yet not being equilateral. This is not so in physical space. A triangle, say on the sand, whose sides are measured carefully and found to be a quarter of a mile in length, can yet yield angles which are not of 60°.

In visual space part of a circle cannot look as if it were straight; in physical space part of a circle can look straight.

What really happens is that in visual and physical space the words have different grammars, and are *not* describing objects from different points of view. If they were we should at once be involved in a science of some sort, and would probably look for a theory which in fact we don't need.

† It is perfectly possible to have movement in visual space without anything changing place in physical space. If one is drunk or dizzy physical objects all give the appearance of altering their place, and yet remain in exactly the same place—visual and physical space. †

It is sometimes said that "This has shape" implies "This has size", and this is regarded as a synthetic *a priori* proposition, and used as an illustration of the fact that there might be synthetic *a priori* propositions of the same kind but of far greater importance.

† The problem is, what does it mean to say "This has shape"? To say shape involves size does not mean that a constant shape involves a constant size (or vice-versa), but that to talk of a constant visual sense-datum means to fix two dimensions, namely shape and size. To talk of "this" in such a context means that it has shape and size. †

Shape and size are merely different ways of describing physical objects and of course are not concepts. What would it be like if shape had not size or size had not shape? They are not entirely distinct ideas; the distinction is made in language.

If I say "This picture has shape", I could mean it in the sense that "This picture has a shape in distinction from this cloud of steam which has no shape". If you say that the cloud of steam has shape at every particular moment, you would mean as distinct from something which changed suddenly.

In both these cases there is no mention of size. Or is it implied that if shape is changing gradually size is also changing gradually?

The right way to put what we are trying to say is that it has sense to ask of any visual sense-datum "What shape has it?" and "What size has it?", but not necessarily to ask this of a physical object, which need not have physical shape entailing physical size, but can have one of them without the other.

A penny placed on the back of one's neck shows that an object can have size without shape (cf. the well-known psychological experiment). Also it is quite possible to imagine a picture frame which, while retaining the same shape, increases and diminishes in size. On the other hand it may remain the same size but be a square at one moment, gradually change to an oblong, then an oval, then a circle. The size can remain constant while the shape varies.

C IV   Auditory and Physical Sounds

Can you in a continuous sound distinguish the part you are hearing at the moment and the part you remember hearing? You can hear a click, and there is no part of it which you can remember as coming before or after another part; whereas with a sustained violin note you can remember the part which has gone before. The problem then is to find an intermediate stage at which you can say that you both hear and remember.

The confusion lies in thinking that physical sound and the sense-datum are both continous. The physical sound is continous, but the sense-datum is not.

The two experiences, hearing and remembering, are quite distinct. You can narrow down the point between where you finish hearing and where you begin remembering, but there will be no point at which you can say you both hear and remember. To say so would be as nonsensical as saying "I see and hear red"—unless you are using either term in an entirely new sense; and there is no need for this as the confusion arises from regarding the continuity of physical object and auditory sense-datum as the same.

Seeing and feeling a pencil is an analogous example. Another is a spectrum of which part is seen, part measured by thermometer, part by an electrical instrument. You can narrow down the point between where you finish using the thermometer and begin to see; but there can be no point at which you can use both sight and thermometer.

C V   Comments on Broad

See Introduction p. xiii. Comments which appear together consecutively in the original notes are indicated by a capital letter at the beginning of the introductory notes below.

A. Broad divided Philosophy into Critical and Speculative (terms which explain themselves) and of Critical Philosophy he said that it had as two of its methods The Principle of the Extreme Case and the Principle of Pickwickian Senses.

The Principle of the Extreme Case he described by saying that if you want to analyse a term it is useful to consider its application to odd and abnormal cases which may make you aware that it is more complex than it appears at first sight.

It is advantageous to point out such cases in order to show where our means of expressing ourselves are ambiguous. But we should not endeavour to include extreme cases in with normal cases or to work a theory to include both.

A. In his Principle of Pickwickian Senses B. distinguished between the common use of a term and its more precise analysis, and said that we can continue to use it in a "Pickwickian sense" even though we know that its precise meaning differs from its common meaning. The reference is to Pickwick Papers *Chapter I, and the examples Broad gives are matter and self, terms which we can continue to use even when we know that their "inner meaning" is very different from their common use.*

This is the very last thing philosophy should do. Broad says that philosophy is endeavouring to get clear; but it is shocking to use words with a meaning they never have in normal life and is the source of much confusion.

A. A third method of Critical Philosophy Broad called Transcendental, which may be characterised briefly as Kant's critical method without the peculiar applications Kant made of it.

This is the right sort of approach. Hume, Descartes and others had tried to start with one proposition such as 'Cogito ergo sum" and work from it to others. Kant disagreed and started with what we know to be so and so, and went on to examine the validity of what we suppose we know.

A. Broad said that Speculative Philosophy had two methods. The deductive which started with certain fundamental self-evident propositions and proceeded to deduce further propositions about reality, and the dialectical which he describes as the Hegelian method of examining contradictions, their relations and resolution.

The deductive method was in effect the same as that of Descartes and others just referred to; the dialectical method is very sound and a way in which we do work. But it should not try to find, from two propositions, a. and b., a further more complex proposition, as Broad's description implied. Its object should be to find out where the ambiguities in our language are.

A. Broad also listed certain objections to philosophy. Briefly, that it reaches no conclusions, that its methods are wrong, that it is taught historically ("and why go on retailing the opinions of back-numbers?"), and that its territory is gradually being taken over by the sciences.

Philosophy does attempt to deal with questions which do not really arise. For example the relation between the penny and its appearances, the relation of body and mind. Once it is clear what the latter question means it is a matter of psychology.

If philosophy were a matter of a choice between rival theories, then it would be sound to teach it historically. But if it is not, then it is a fault to teach it historically, because it is quite unnecessary; we can tackle the subject direct, without any need to consider history.

B. Broad suggested that there were three "theories of truth", the Correspondence Theory, according to which truth is the correspondence of judgement and fact, the Coherence Theory, according to which truth is coherence with some system of other judgements, and the Pragmatic Theory, according to which truth is what works.

Philosophy is not a choice between different "theories". It is wrong to say that there is any one theory of truth, for truth is not a concept. We can say that the word has at least three meanings; but it is mistaken to assume that any one of these theories can give the whole grammar of how we use the word, or to endeavour to fit into a single theory cases which do not seem to agree with it.

Of the three theories, W. made no comment on the Correspondence Theory, beyond giving "George V is King of England" as an example of the correspondence of judgment and fact, but commented as follows on the Coherence and Pragmatic theories.

Coherence. In a court of law statements are often taken as true if they are coherent with the rest of the facts, even though it may not be possible to verify them.

Pragmatism. The hypothesis that there are electrons is taken as being true because in practice you can work as if it were the case. So also Einstein's Theory of Relativity is accepted because it works in practice. Thus Euclidean space is used for everyday purposes, and relativity for immeasurable and astronomical distances. To decide between them would need a great deal of empirical evidence, and this is certainly the sense of truth we apply to them.

But we do also use the word true of a ruler, which is a sense not ascribable to any of the other examples given.

Thus it is nonsense to try to find *a* theory of truth, because we can see that in everyday life we use the word quite clearly and definitely in these different senses.

B. Broad also put forward an "Epistemological Classification of Judgements", in which he distinguished between a priori *judgements and empirical judgements.*

A priori and empirical are not two kinds of proposition. What is there in common between them that we should call them "propositions"? An *a priori* proposition would have to be one whose meaning guaranteed its truth. But meaning requires us to verify.

Expressions which look like *a priori* propositions must be clarified. Just as the same expression can be a proposition or an hypothesis, so the same expression can be an equation or an hypothesis. Unless we distinguish confusions occur.

An equation is necessary, and a rule of grammar; and is therefore arbitrary. Geometry is the grammar of certain expressions—cube, circle, point, line etc. Arithmetic is the grammar of $1,2,3,4----,+,=$ etc.

People thought that there might be *a priori* metaphysics seeing that there was *a priori* mathematics.

A priori propositions as they are used in traditional philosophy look like an anticipation of experience.

Propositions usually taken to be *a priori* and empirical are

(1) *A priori.* $2+2=4$, All equilateral triangles are equiangular.

(2) Empirical. The kettle is in the grate.

As has just been said, the same expression can be an hypothesis or a proposition in the strict sense. So also a proposition can be *a priori* or empirical. Thus to take 2+2=4, as an example, it has sense to say of visual space that two objects + two other objects make four objects. And 5+3 might be seen in visual space to be 8. And this is *a priori*, no empirical proof is needed. But in physical space 2+2=4 is an hypothesis and needs verification. If you see two objects in one room and then two objects in another room, it is an hypothesis to say that they make four, because when you come to put them together they may make five objects. This can never happen in the visual field. You could not see four drops of rain-water which were in two groups as anything but four; whereas in the physical world they may coalesce to make one large drop.

It has sense to say that all equilateral triangles in visual space are equiangular, because you can never see the one characteristic without the other. The proposition is therefore *a priori*. On the other hand in physical space you can easily imagine an experiment on sand where an equilateral triangle is drawn whose angles when measured were not equal. Some propositions are *a priori* in visual space but in physical space empirical. The same proposition can be either.

"The kettle is in the grate", however, is entirely empirical, because it needs a particular type of verification.

Compare the experiment in psychology in which you are presented with three objects. You do not have to *count* the objects—you see them as three. And there is no further verification possible as this is a proposition about your sense-data. On the other hand if you are shown thirteen objects, you cannot in the same sense say that you see thirteen. You need to verify. You might count them and find there were fifteen.

C. Broad distinguished between inspective judgements (this looks red), perceptual judgements (this rose is red) and inductive judgements (all swans are white).

The inspective judgement is concerned with sense data, the perceptual is a hypothesis which may be true or false, and the inductive again a hypothesis which may be true or false, but which is sometimes said to be less true than the perceptual judgement.

C. Broad distinguished different degrees of certainty.

There is a psychological and a logical sense of certainty. If we say "This chair is green" we are saying something psychological and empirical, which is verifiable in a different sense from "This chair looks green" where our certainty is logical and *a priori*.

C. Broad distinguished a priori and empirical concepts, and gave "ideal limits" (the perfect circle etc) and categories (cause and other "very general types of order") as the only "plausible" examples of the former.

What does it mean to hold that there are *a priori* concepts? If we pull a piece of cotton very tight, then to say that it is straight is to refer to what is manifest to our senses; it is a judgement about our sense data. But we know perfectly well that if we look through a magnifying glass we shall see that what was apparently straight actually is not so.

This shows that there can be appearance and reality in a certain specific sense, which can be explained, generally, by the distinction between visual and physical space. We know this to be so with a line or circle drawn on paper; visually they look perfectly straight or round but physically they are not so. In imagination we sometimes imagine a perfectly round circle, for example. But this is not a hypothesis which can be verified.

If cause were an *a priori* category you could make causal propositions. But cause is always a hypothesis and in this sense there are no causal propositions.

You can say "The fire causes the kettle to boil", and yet when you come to try it the kettle might fail to boil. "Thunder comes after lightning" is a proposition, being a statement about sense-data; "Lightning causes thunder" is an hypothesis and needs verification and may be true or false.

Causal necessity. The laws of nature are not outside phenomena. They are a part of language and of our way of describing things; you cannot discuss them apart from their physical manifestation.

C. Broad distinguished various "theories of knowledge".
Two of them were Rationalism, which holds that there are
some synthetic a priori *judgements and* a priori *concepts,*
and Empiricism, which holds that there are neither.

The rationalists were right in seeing that philosophy was not empirical, that is, that as soon as it became empirical it became a question for a science of some sort.

But they were wrong in supposing that there were *a priori* synthetic judgements. They endeavoured to settle everything by reason, by sitting in arm-chairs and inspecting words— they let the words speak to them.

The empiricists saw that we could only describe the world. They missed the point when they tried to make philosophy empirical, but they were right in maintaining that reason could not settle everything, and that synthetic propositions were matters of experience.

D. Broad also spoke of scepticism, a sceptic being one who "doubts the truth of all judgements".

If someone argues that you only came into existence ten minutes ago, as Russell suggested, he is using words in a new sense. Yet he means[1] exactly what we mean; cf Descartes' malicious demon.

(Russell, *Outline of Philosophy* p 7: Moore, "Wittgenstein's Lectures", *Mind* 1955 p 25: "part of what we are saying when we say that an event in happening 'now', is that it was preceded by other events which we remember")

Idealists were right in that we never transcend experience. Mind and matter is a division *in* experience. Realists were right in protesting that chairs do exist. They get into trouble because they think that sense-data and physical objects are causally related.

Idealists saw that a hypothesis was not something outside experience. Realists saw that a hypothesis was not merely a proposition about experience.

That a tennis ball looks round is a proposition. But that the earth is round is only a hypothesis. There is no fact that

[1] Perhaps 'thinks he means' would bring out the meaning better.

the earth is round over and above the various facts such as the shape of the shadow on the moon at an eclipse, ships disappearing over the horizon etc; just as there is no fact that this is a physical object over and above the qualities and judgements of sense-data about it.

There is a tendency to make the relation between physical objects and sense-data a contingent relation. Hence such phrases as 'caused by', 'beyond', 'outside'. But the world is not composed of sense-data and physical objects. The relation between them is one in language—a necessary relation. If there were a relation of causation, you could ask whether anyone has ever seen a physical object causing sense-data. We can talk about the same object in terms either of sense-data or hypothesis.

C VI Sense-data

Sense-data are the source of our concepts: they are not caused by our concepts.

In the primary sense you do not see with your eyes; the correlation is contingent. You see what you are dreaming, but not with your eyes.

All causal laws are learned by experience. We cannot therefore learn what is the cause of experience. If you give a scientific explanation of what happens, for instance, when you see, you are again describing an experience.

All propositions about causation are learned from sense-data. Therefore no proposition can be about the cause of sense-data.

You can't say that a cyclone causes this sort of weather; because to say there is a cyclone is to say that there is this sort of weather. You can't say my knowledge of evolution is the result of evolution.

The world we live in is the world of sense-data; but the world we talk about is the world of physical objects.

C VII Certainty and Relevance

"I am sure" is either a grammatical expression or a statement or proposition. An example of the first of these two uses[2] is "I am sure that $2 \times 2 = 4$". But here it really has no sense to say "I am sure"; for wherever "I am sure" is followed by a proposition it is unnecessary, and equivalent to the multiplier 1 (e.g. $276 \times 1 = 276$).

Where "I am sure", "I can be sure" or "I cannot be sure" are statements or propositions, then you can say that you are sure that someone is in the next room, and yet it may be false. Similarly with "know", you can say that you know something and yet it may not be true.

If you say "It seems to me that B. has a red tie", this cannot be disputed. "It seems to me" is ultimate and a proposition. But "B. has a red tie" can be disputed. To be sure of a proposition means simply "It seems to me".

If there is no test to decide between two statements then they mean the same. Galileo's correct answer to the prelate who refused to look through his telescope, because he said that whatever he saw would be the work of the Devil and false, should have been "I quite agree with you; we both mean exactly the same thing".

If you always cheat, you can never say you cheat.

[2]The relation between the two alternatives is not quite clear in the original and I have amended slightly to give what seems to be the intended sense.

You can't lay the foundations of language before they are there.

"Because" and "why" can refer either to a reason or a cause. If a red traffic signal acts on you in a manner analogous to a drug, then your explanation of your action is giving a cause. If on the other hand you see the red light and act as if someone had said "The red light means stop" then your explanation would be giving a reason.

Reference to past experience must be reference to propositions about the past.

If you remember wrongly, then there must be some criterion other than your remembering. If you admit another test, then your memory itself is not the test.

If you keep on asking "How do you remember, know etc?" you will ultimately be driven to saying "It seems to me". If you say you remember and that is your only criterion, then you cannot go beyond it.

We don't start with hypotheses and then go on to propositions. A hypothesis is a law by which we can construct propositions. The use of a hypothesis is to make inferences about the future.

C VIII   Games and Rules: I

We can only describe what happens, but we often talk about *rules*. How do we follow a rule?

(1) We can find out the rules of a game in the way in which

a scientist investigates chemical or physical laws. We can observe how the game is played.

(2) We can ask the players, and what we find out is then quite different. They give us reasons and rules. (An observer might find out some rules of which the players were not aware.)

If then we say that the game is played "according to the rules" we mean either (1) compatibly with the rules or (2) we regard the rules as motives, and when we give the rules we give the reasons. (It is false to suppose that there is always a reason but that you do not know it.)

What we investigate is one particular game or another, not games in general or something metalogical. We need not recapitulate the rule as we play—we use words without looking them up. If you give the rule you are doing all you can. Rules are fixed and given: they allow some combinations and disallow others.

If we played a game of chess and made only three moves, we could legitimately be asked "Were you playing chess?" as the same moves might have been common to ten different games.

We are led to believe that intention and understanding accompany a proposition as a toothache might. But if you say you intended to play chess, you can say when you started and ceased. To say "I thought of this for twenty minutes" does not mean you had a feeling. If it did you could ask "Did you repeat the phrase (e.g.: I expected Mr. Smith to come into the room) all the time?"

If you describe a picture the picture is not in every word.

A picture corresponds to a proposition.

Thinking means operating with plans.

Language is not an indirect way of communicating what thought-reading could communicate directly. The same is true of visual imagery.

We cannot say that an expectation is similar to what is expected, because similarity requires comparison, and what is expected is not yet available for comparison. Things are said to be similar according to some rule of projection; but some rules of projection are more familiar than others.

C IX   Symbol and Thought

A symbol does not work by suggestive powers. The meaning of a word is not the object corresponding to it, but the grammatical rules which apply to it.

A proposition is a fact which is a picture of another fact.

You can describe the experience of learning a particular language, but you can't describe the experience of learning to use language because you would then have to be able to think what it was like to have no language at all—i.e. to think what it would be like not to think.

At the primary level experience hasn't a beginning. Visual

image, sentences on paper, spoken sentence, picture—all are on exactly the same level.

We can't talk about the limit of experience, because we should have to experience both sides of the limit.

In the process of thinking, the thought does not appear first, to be translated subsequently by us into words or other symbols. There is not something which exists before it's put into words or imagery.

C X Grammatical Rules Arbitrary

We use arbitrary or not arbitrary with reference to a rule. If you ask why we adopt the rule and whether it is arbitrary, it depends. Either it is just arbitrary, or else we say it is the simplest rule to adopt.

A grammatical rule does not stand in a relation to reality such that we can give rule and reality and then see whether they agree or not (cf. the gulf between picture and reality). We don't model the grammar of "red", "green" and other such words to match reality, because this would mean that we could say "these qualities have this sort of grammar". The grammatical rules applying to "red", "green" etc cannot be justified by anything you can say about colours. If you could justify them, they would not be arbitrary. You can't justify the grammar of "red", "green" by anything you can say about them, because in using these words you are already using a grammar and language.

You can't justify "not" by saying it is a sign for negation because in doing so you are using negation.

You can only get hold of the meaning of a word through the rules with which you use the word. We say "Twice two is four" and "This is red". But when we point to a rose and say this, do we mean that it is identical with red? Of course not. All we can do is to show that the two signs are used differently; and the person to whom we are explaining can only be satisfied. This is all we can do.

† If you say that the rules of grammar are arbitrary, you perhaps expect some further set of rules to justify them. But these rules will then in turn need justification. Grammar is self-contained. Talking nonsense is not following the rules.

Grammar is not something higher, with another grammar beyond it. And if you say "We must follow grammar if ---", this leads to the question "Why do we use grammar?" But we are not concerned with what would happen if we were *not* playing this game; and this is the justification of grammar.

"What would happen if we didn't follow grammar, of if our grammar was not what it is?" The whole point is that if we did not have our grammar, we should have to pass to another grammar. †

C XI   Why do we think at all?

If a man struggles against having his hand put in the fire, you can ask "Why does he struggle against it when he hasn't been burnt yet?" Was the motive something which accompanied the struggle?

No process of thought goes on continuously. We could say that he has good reason to kick if his reason is that he has been burnt before. But he still may or may not be burnt this time. There is no justification yet.

If he says he "is likely" to be burnt, this seems to be a new suggestion. The reason why it's likely is that he has been burnt a thousand times before. But we can omit the "likely"; for whatever reason he gives, he still may be burnt or not.

Evidence is always in the past. A general proposition is justified by the reasons we can give, not by the results.

However far the reasons go, they stop short before the fact. Trains of reasoning go on and then something either happens or does not.

We can't talk of reasons for thinking. We can't say "We must think because - - - -". We can describe the game of thinking, but not the reasons why we think. "Reason" only applies within a system of rules.

Thus if you ask "What are the reasons for thinking?" all you can give is the rules. The justification for inference is a rule.

It is nonsense to ask for reasons for the whole system of thought. You cannot give justification for the rules.

The idea of laying down a rule is that if we are in a philosophic difficulty, the rule will decide for us. If we were discussing Moses we should not feel the need of a definition. But if something cropped up where there were two alternatives, some particular rule would be needed to settle it.

We lay down lists of rules, and don't always decide which we are using.

What role do the rules play? When we are dealing with actual misunderstandings, we need to agree on a definite rule. If a word has three meanings, and we admit that there are three different definitions by description of the way in which people use it, an exact rule is needed when it is a question of avoiding a misunderstanding.

How can we decide which game we are playing? In most cases we don't decide. We constantly discuss tables of rules, but for the reasons just given we hardly ever use them explicitly.

C XII "All", "every", "and so on".

If you infer from "The whole door is black" that "This part of the door is black", does this mean that "The whole door is black" is a logical product? It certainly does not mean that all the different *parts* are black; it refers to *visual* space. For it is conceivable that the physical object might be divided up into a number of black spots.

We constantly make the error of muddling up the visual field with a painted picture.

"All" and "any" have many different meanings, just as 1,2,3 can be cardinal, ordinal or rationa; though there are formal similarities in the use of the words.

Where you explain by giving instances and saying "and so on", this is a perfectly good explanation. But "and so on" means different things in different cases. It *can* have a strict and exact grammar. But when we write "- - - -" we normally regard this as essentially inexact. It is a notation like all others if you know how to use it; it seems to suggest some things distinctly and then merge into a fog.

a) $1^2+2^2+3^2 - - - = 1^2+2^2+3^2+4^2 - - -$

b) The number of circles which you can draw between two lines.

a) and b) are different. When you are asked how many circles can be drawn between two lines, it is nonsense to expect a cardinal number as an answer, and you cannot answer by giving one.

Grammar will vary according to whether the reference is to the visual field or to the Euclidean. Identity of length is entirely different in visual space and Euclidean space.

a,b,c - - -, and so on. Here - - - or "and so on" stands for the rest of the alphabet, a definite number of letters.

This is quite different from $\frac{1}{3}=0.33$ - - - and so on. Here there is no definite number of digits, nor could there be for some superior being. The two examples have different grammars and rules. 0.33 - - - is not a makeshift: it has an exact grammar.

a) $(x)\psi x=\psi a.\psi b.\psi c$--
b) $(\exists x)\,\psi x=\psi a\text{ v }\psi b\text{ v }\psi c$---

Both a) and b) are a translation from our language: they do not remove the ambiguities of our use of "some" or "all" in language.

In chemical analysis you investigate and find: but not in logical analysis.

What exactly we mean by a heap of sand is in a sense vague, because if you put grains of sand down one by one, you can ask when do they become a heap. If the government were to say that 5000 grains of sand make a heap, that would not fix our ideas in the past; it would make a new rule. The ordinary pace has no fixed length; you can say it is between a yard and a yard and a half. But if you fix it you are not making one figure right and the other wrong; both *were* right, but the calculus is now different.

A mathematical proof is not illustrated by a figure in the same way that Alice in Wonderland is illustrated. In mathematics proof and figure are the same thing in different symbols.

C XIII   Calculation

"To follow rules" is an ambiguous expression. 100 and 101 dashes on the blackboard are, in visual space, exactly alike. Counting is a different game from what we do in visual space, and different again from what we do when we see *three* dashes.

Uncertainty in a calculus does not mean that there is certainty beyond. It simply indicates *another* calculus. Whatever is laid down to be the test is the test.

If you multiply two numbers together you get a result, but you can do the same multiplication again and get a different result, even though both sums may be done according to the same calculus. The product you get is in fact a hypothesis; e.g. $14 \times 14 = 196$. But $2 \times 2 = 4$ is not a hypothesis, and the calculus is different.

In visual space there are at least three calculi,

(1) Where you look and can tell at a glance e.g. XXX.

(2) Where you look and see that nothing has changed: e.g.

$$\begin{matrix} \bullet & \bullet \\ \bullet & \bullet \\ \bullet & \bullet \end{matrix}$$

(3) Where you look and can't tell whether there are any changes.

These are all different.

†You cannot see a hundred dots: you can only see many.†

Where does calculation take place?

It can take place either on paper or in visual space, but the game is different in each case. So you could do your calculation on a piece of paper or on a large scale in Trinity Great Court; but the two calculations would be different, and you could be surer of the one than the other.

There are different ways of ascertaining that numbers are equal. Compare the one to one correlation in the figure below--

You could see these two rows as having the same number of crosses each. But if you came to count them you might find a difference between them. Which is correct?

The result of a calculation is not a matter of experience, but that which we accept. $20 \times 20 = 400$ is a rule of grammar. We can provide such a grammar in two ways. (1) We can teach you to multiply; (2) We can supply you with multiplication tables already worked out. Any uncertainty which adheres is one which adheres to *all* grammars; but it does not stop it from being grammar. (Fixing the length of the metre is fixing grammar.) The uncertainty of multiplication is one which adheres to the whole of grammar. But that does not mean that the rules of grammar are experimental propositions.

Are words like grammar, language, proposition, rule, calculus, mathematics, logic and so on on a different level from others? We discuss these in philosophy, but not words like table, chair and so on. Are the second type on a different level? No! Language as opposed to what? Washing?

We are inclined to say that the rules of grammar *must* be what they are. What if they weren't?

Contradiction is between one rule and another, not between rule and reality. One feels that a wrong rule is an obstruction in the way, but in fact it is only another different way. If you give the *wrong* rule, you give a *different* rule. If you feel unable to get over the obstruction, it is not in the physical sense 'unable'. In the physical sense you can try to remove the obstruction. In the other sense it is a choice between one or other calculus.

We often want to say something like, "If I want to express *this,* then I can't use *these* rules". But how did you fix the rules for "this"? You are not talking about a natural object, to which you are pointing and saying you want to copy it. If you say "I want to describe this fact" then we can ask "What fact?"

There are no necessary facts; all facts are contingent. But this is not a classification like "All lions have tails", which is a matter of experience and could be otherwise. No fact can be necessary, for if it has sense to affirm it, it also has sense to negate it. If there was not this choice it would have no meaning to make the assertion. The negation of a proposition must have sense. But if there were a necessary fact, it could not be otherwise.

When we say this it sounds as if we were stating a proposition, because we are inclined to think that a proposition works by some sort of suggestive powers and that there is no reason to believe that the negation of a proposition also symbolises something. But the proposition does not work by suggestion, but as a picture or projection of reality; and since a fact can either be a fact or not, there must be the same possibility in the picture or projection.

There are no positive or negative facts. "Positive" and "negative" refer to the form of propositions, and not to the facts which verify or falsify them. A negative statement does not have meaning in the same way as a positive one; you cannot describe it in positive terms and retain its negative meaning.

C XIV   Rules and Grammar

What does it mean to say "I can't imagine red and blue in the same spot at the same time"? It sounds as if it were a statement about psychology. But in fact it is nonsense to suggest we can even try to imagine it. It is impossible, though not in the sense in which we say that it is impossible to lift a man with one hand. When you say you can't imagine red and blue together on the same spot, this is a rule of grammar.

To "know" something is not one clear-cut psychical event. How long does it take to know a thing? How long does it take you to know how to play chess? There is no parallel here with "How long does a piece of music last?" The ability, the knowledge how to play chess has no temporal structure; the game of chess has. But language misleads us into thinking that knowledge has duration like a toothache or a melody. But when we say we know something "all the time" this is a quite different sense. It is a hypothesis, and we can correlate it with a physiological process.

How could we explain a paradigm from which we derived grammar? If you have a paradigm, you again have grammar; and if you assert something of one word and deny it of another you are already using grammar. Logicians try to catch grammar unawares; but it's like trying to catch your own thumb.

The meaning of a word is given by the explanation of that

word. The criterion of understanding is that you can give the explanation before you know whether a proposition is true or false. Eating breakfast is not a paradigm for the grammar of eating breakfast; it simply enables you to say whether you are eating your breakfast or not.

No fact can be a paradigm for grammar. If you try to find facts to justify grammar, when you try to *say* what they are, they are no longer paradigms; you are using the same grammar to describe both the facts and the grammar you are justifying. Can we say "If we had a different grammar we should get into conflict with the facts"? Changing your grammar can never do this, because if you say that so and so is the case you can also negate it. Grammar can never get us into this sort of trouble of saying something which is not true.

Grammar is not determined by facts. You can only get into conflict with reality by saying something which is not true. We cannot describe a paradigm for grammar because we should have to use language to do it.

C XV   On Mathematics

In mathematics there are many different calculi: visual, physical, definition, rule of grammar, hypothesis and proposition.

For instance A and B in the table on p. 96 are entirely different as calculations. A may be a proposition for a developed human being, because he may know at a glance that there are 16 crosses. B on the other hand is a hypothesis; it would have to be counted and the fact that there were 16 crosses verified. C and D are also propositions. That they are five and three are propositions about visual space; no counting is necessary.

E may also be a proposition; to an adult it needs no verification. The multiplication tables on the other hand are definitions, and the number of tables which are definitions varies with each individual. For most people the tables up the twelve times table are definitions. But other instances may also be definitions. Thus for me $15 \times 15 = 225$ is a definition, while it might be a hypothesis for someone else. $7^3 = 343$ is also a definition for me. But $16 \times 16 = x$ is for me a hypothesis until I can work it out. But if after working it out I simply know that $16 \times 16 = 256$ then it has become a definition.

```
A  X X X X    B      X        C  X   X     D  X           X
   X X X X         X               X
   X X X X       X                 X   X
   X X X X           X X X X
                   X X X X X
                     X X X
                       X                F   123
                                            753
                E  11                       369 x
                   11                      6615 y
                   11                       861 z
                   11                      92619
                  121
```

It would have no sense to say of A, "I see 16 crosses and yet when I count them there are only 15." But if there were two extremely long rows of dots in one to one correlation, with say 100 in each row, it would have good sense to say "I see these rows as being equal, yet when I count them one has 100 dots in it whereas the other 101". Here we have a hypothesis, whereas A is a proposition.

F is a rule of grammar, or a calculus done on paper; but individual parts of the work may be done according to one of the calculi mentioned. So step x is for me a definition; step y is a hypothesis, but the first stage in it, $5 \times 3 = 15$, is again a definition. The result is a hypothesis. Someone else may go through the calculation and get a different result. The individual steps are rules of grammar and the whole process is a rule of grammar.

C XVI   Moore's Criticism

In his articles on Wittgenstein's lectures in Mind *Moore wrote (*Mind *1954, p. 298) "I wrote a short paper for him in which I said that I did not understand how he was using the expression 'rules of grammar' and gave reasons for thinking he was not using it in its ordinary sense; but he, though he expressed approval of my paper, insisted at that time that he was using the expression in its ordinary sense." Moore dates this paper after the academic year 1930–1931, and the entry in J.E.K.'s notes at this point therefore seems to refer to it.*

Moore read a brief paper to introduce a discussion on Rules of Grammar. He gave two examples.

(1) Where there is no doubt. "Three men was working". Here it is clear what the rule is and how it has been broken.

(2) "Different colours cannot be in the same place in a visual field at the same time." This differs from example (1). Are the two examples rules of grammar in the same sense? If we say "Two colours can't be in the same place", we may mean that we can't imagine it, that it is inconceivable or unthinkable, or that it is logically (as distinct from physically) impossible.

W.'s reply.

The right expression is "It does not have sense to say - - -"; but we usually express it badly by speaking of a rule of grammar. So it does not have sense to say "This table is as identical as the other". Compare using the same board and the same pieces as we use for chess, but making moves which the rules do not provide for.

We have a feeling that the first misuse referred to by Moore is harmless but the second vicious. But in fact both kinds of rule are rules in the same sense. It is just that some have been the subject of philosophical discussion and some have not. If we discuss a rule we have to state it.

These difficulties arise from false analogy. So the puzzle that we "can't" measure time is due to the analogy of the physical "can't". We are inclined to say that we *can't* imagine or think something, and imply that we *could* express it correctly if we had the experience. To say that something is "logically impossible" sounds like a proposition. So if we say we can't think of red and blue together in the same visual space, we have a feeling of *trying* to do so, as if we were talking about the physical world; we somehow cheat ourselves and think it *can* be done.

Grammatical rules are all of the same kind, but it is not the same mistake if a man breaks one as if he breaks another. If he uses "was" instead of "were" it causes no confusion; but in the other example the analogy with physical space (c.f. two people in the *same* chair) does cause confusion. When we say we can't think of two colours in the same place, we make the mistake of thinking this is a proposition, though it is not; and we would never try to say it if we were not misled by an analogy. It is misleading to use the word "can't" because it suggests a wrong analogy. We should say, "It has no sense to say - - -".

The rule about red and blue ((2) above) is a rule about the use of the word 'and'; and we would only say that 'was' ((1) above) makes nonsense if someone said it posed a philosophical problem.

C XVII   Hidden Logical Products and Unsolved Mathematical Problems.

"Hidden" here is not used in the sense in which we say a man who is in another room is hidden. What is the criterion of discovery? If there is none, then you can't tell whether you have discovered or not; you must have some method of discovering the hidden logical product. If you have no method then there is no sense in calling it hidden.

This leads to the question of unsolved mathematical problems.

You cannot tell whether in the evaluation of π three sevens ever occur consecutively. If you divide a circle into five equal parts by luck, and then try to find a construction to do it, if you succeed you have in a sense not found what you were looking for. When you divided with pen or pencil, you had done it; but the construction gives you something new or different.

Geometrical lines, points, etc. are sometimes spoken of as if they were more perfect than physical ones, and traced so to speak with a real pencil. Nothing of the sort is the case. The only circles we know are real ones, not "geometrical" ones.

Your division in physical space may not divide the circle into five exactly equal parts. Does the construction produce a regular pentagon in physical space? One is tempted to say "Yes, if correctly done". But what is the relation between the construction and the physical pentagon? If the two coincide it proves nothing because you may make a mistake in either. If it has sense to say that the more exact the drawing the better the agreement between the two results, this can only be tested in physical space by experience.

The notion of a regular pentagon was not that of a constructed pentagon, nor was it something that could be looked for. When the construction of a regular pentagon was discovered, a new symbolism was created; this had a relation to the old one in that by means of experimental propositions about physical space we can now use the construction as a kind of measure for the physical object.

You know what it means to divide an angle into three equal parts. When it was proved that there was no construction which would do this, it showed that the efforts to find one had been futile, and not the efforts that they were thought to be.

Every problem in mathematics is exactly of this nature; the result is never what is asked for. Construction is only the road to a given goal; if you find a construction it will give you something new.

In unsolved mathematical problems, the question asks for something different from the answer reached. The problem is set in language which does not state its desired solution, but something analogous to it. If a solution can be found, the proof does not prove what was stated but something else.

We can't investigate the game of chess, because we have made it up. If we make up another game like it, the result is a new game. The old game remains what it is.

What can't be found in logic and mathematics is not hidden.

C XVIII Games and Rules: II

If you lay down rules for a game you don't really *follow* them when you are playing. You don't play chess with continual reference to the rules; they are there to be referred to if you get stuck. You don't even have to remember the rules; compare what happens when you look them up.

Is playing chess then an amorphous state like toothache? You can't say "we may either repeat the rules or act like a machine". It's quite different from that. You can't say that *either* on the one hand you are a parrot *or* you look up the rules. It is infinitely more complicated. Why do you call primitive games without rules games?

If you are expecting someone to tea there are endless ways of showing it—thinking, preparing, saying so, and so on.

Is there any difference between expecting a man to tea at 4.30 and expecting him at 4.40? The difference may not show itself at all unless you ask or look it up in your diary; and the criterion might be the asking or the looking up. Your excitement about Smith coming at 4.30, your belief or fear or hope are not in your head at all. The criterion is what you say you expect. (In actual fact you have no one criterion.)

C XIX Language and Reality

Using the word "language" is dangerous; but no more dangerous than using the expression "word game". You still can't decide in borderline cases whether they are "languages" or "games".

If we say (a) Red is *rot* (b) Red is the colour of this patch (c) 2×2 is 4, then—

(c) is different from (b);
looking for a red flower is different from looking for a flower of the colour of this patch;
(a) is different from (b);
but (a) is similar to (c).

One feels like saying that (a) is in the realm of language only, but that (b) gives a connection between language and reality; and we generally draw the line between language and reality at the written and spoken word.

If you send a sample to a tailor it may or may not be part of language. The connection is made only by rules. A red patch can also be a part of a language. But we agree that the use of a sample is different from the use of a word.

One of the implements of our language is ostensive definition. But with such ostensive signs we have only a mere calculus.

What we call a connection between language and reality is the connection between spoken language and, for example, the language of gestures. If we had no written or spoken language, where then would be the connection? How can you explain one gesture by another?

Language reaches into reality. (b) above makes a connection between one part of language and another, and (a) makes a connection between letters and sounds. We make these connections. And we speak of our written and spoken signs as "language", but not our samples; though in a sense a sample *is* part of language.

Russell found in our language relations which were connected in a calculus. But if a man uses a calculus of three-term relations and gives jealousy as an example, he has not given it any more substance. He has applied the calculus of three-term relations to jealousy; i.e. our calculus in ordinary life is partly the same as his. But this is not an application in the sense in which it is an application of a calculus if you build a model of a bridge. R (abc) is an application of a three-term relation. Compare four-dimensional geometry when someone finds time as the fourth. He has not found something new, and can't say "Now I have got it". He has it already in language. It has all the reality it will ever have; it can be connected up with the larger calculi of our language. You can't rest it on anything; the calculus does not rest on reality.

We have the idea that language is kept in bounds by reality, or by the connection with reality, in the way in which the motion of the planets controls the falsehood and truth of our statements about them.

C XX   Physics and Causality

Physicists make reference to laws of causality in a preface, but they never mention them again. They can't deduce their axioms from causality, but think it may be done some time. There is nothing extraordinary in this, but they never really dream of causality in this sense. Yet, in a different way, causality is at the bottom of what they do. It is really a description of the style of their investigation. Causality stands with the physicist for a style of thinking. Compare in religion the postulate of a creator. In a sense it seems to be an explanation, yet in another it does not explain at all. Compare a workman who finishes something off with a spiral. He can do it so that it ends in a knob or tapers off to a point. So with creation. God is one style; the nebula another. A style gives us satisfaction; but one style is not more rational than another. Remarks about science have nothing to do with the progress of science. They rather are a style, which gives satisfaction. "Rational" is a word whose use is similar.

Is induction justifiable? Someone might say "Oh, it must be, because it makes things more probable". But again it is like the knob or tapering of the spiral. We try to cover up the beginning of our reasoning; but actually reasoning never started. To say that the earth is not supported is exactly similar to saying that language is not supported (though we only speak of supporting where the earth is concerned, and not elsewhere). The whole chain of our reasoning is not supported any more than the earth is supported; the whole of grammar is not supported in the sense that a sentence is supported by reality. You can't in fact call language or grammar unsupported because there is no question of its being supported. But the rules of grammar are not a deduction from the nature of reality.

Someone said, "Once the ostensive definition of red is made, everything is fixed", in the sense that the other rules follow. But compare the word "pal". You can explain it in a language already known; but an ostensive definition can convey nothing. We already have rules for the ostensive definition of red. If we say "This patch is red" we already know that it is the colour and not the shape that is meant. We think the other rules follow because we already know them.

Language is not tied down; but one part is tied to another. To note down a date is rational because it connects up with all experience. Can you ask whether the whole is rational? No: rational applies only within the system. If anyone asked you why you noted the date, you would answer "My memory is bad." And you always give a reason of this type. You can give a finite chain of reasons for calculating the bursting-point of a boiler. But if someone goes on asking for reasons you can't give them a further one. All you can say is "I won't build a boiler without making these calculations". If you give a causal reason, then it is not a real reason.

C XXI   Hidden Contradictions and Constructions

The use of the word "hidden" in mathematics is not the same as its use to describe, say, a piece of chalk for which you are looking. Hidden, looking and finding have entirely different meanings in mathematics from their meanings in ordinary language. An expedition to the pole can be described in detail in advance; but where there is no method what are we guided by?

The problems of physics and of mathematics are entirely different.
   (i) 321×47 (which is not really a problem)
   (ii) The specific weight of helium
   (iii) Proof of a theorem
These are all entirely different problems.

When we look for a contradiction we have an idea of what we are looking for. Compare the construction of a pentagon. In a sense one knows what it is and what is wanted—we know what a regular pentagon is. We haven't yet got a precise description, yet in another sense we know what we are looking for. Our description gives us part but not the whole of what we want. For the constructed pentagon is not simply the regular pentagon for which we are looking. We don't *look for* anything; we *construct* something. In the process we are guided by something almost aesthetic. Mathematicians have no method; they seem to think mathematics is like an illness.

Trying to find Mr. Smith and trying to find the construction of a regular pentagon are entirely different. The latter has no meaning. To explain the word "construct" you would have to give examples.

If we look for a hidden contradiction but don't quite know how to find it, we are vague, but have some idea of the form p. ∼ p. So also you can say that the end of a game of chess is the taking of the king, but there are a hundred ways in which you can do it.

A man who looks for a hidden contradiction does not really know the game; he is not doing what he says he is doing. With insoluble mathematical theorems, the proof of the theorem alters the theorem; proof connects it up with something with which it was not connected before.

As soon as we have defined a hidden contradiction we have found it.

C XXII Infinity

Will three consecutive sevens ever occur in an evaluation of π? People have an idea that this is a problem because they think that if we knew the whole evaluation we should know, and the fact that we don't know is merely a human weakness. This is a subterfuge. The mistake lies in the misuse of the word infinite, which is not the name of a numeral.

"If we find that three consecutive sevens occur, then we have proved that they do; but if we don't find them we still have not proved that they do not". This gives us no criterion for falsehood, but only for truth.

It is not that you don't know a way; it is that you have not provided a criterion. When we have described our number, we have a number—it is not something else apart from the description.

A proposition for which there is no criterion of its falsity but only of its truth (or vice-versa) is quite different from one for which both alternatives are provided.

We say "the world will come to an end some day", and in using "some day" we think or feel that we have grasped the infinite. We know how we normally use "some day", "the day after tomorrow", "next week" and so on. But when we say "the world will come to an end some day" what we give is not a disjunction but an indefinite "and so on". There is infinite possibility, but no infinite reality—so we often feel. The infinite does not stand for a number or quantity.

Compare a ruler with infinite radius of curvature, i.e. straight.

If you give a promise to provide any amount of cash asked, your promise is infinite.

"Proof" and "proposition" in mathematics are used in a number of different senses.

You can prove that the angle in this semi-circle is a right angle. If you then draw another similar triangle, you assume that the proof applies in this second case also and in any other case.

I can divide a length by tossing a coin; if heads I bisect the right-hand interval, if tails the left. In visual space the freedom to do this is restricted; in Euclidean not at all, the choice is infinite.

You can ask what is the number of men or chairs, etc. in a group. You know how to get to know the number. But you cannot ask what is the number of cardinal numbers, or prime numbers or points on a line. What would you do to find out? There is no number of them which God knows and we endeavour to find. There is only the general law which leads to their production.

If one is divided by three there is no such thing as an infinite series of threes. There is a law that one divided by three is 0.3 recurring. We confuse the infinite possibility of writing threes with threes written down. Similarly, there is an infinite possibility of constructing points on a line, but a line is not therefore made up of points.

We find an analogy, embody it in our language and then can't see where it ceases to hold.

MISCELLANEOUS NOTES[1]

Change is *either* perceptible or observable continuity of motion *or* succession: the two are logically different.

Analysis is used in two different senses. (1) To mean Definition: statements of equivalence unfolding meaning already known. Any definition in this sense which gives more or less than the meaning being defined is a bad definition. (2) To mean description, as e.g. in chemical analysis, an empirical process.

Certainty. A thing can only be said to be certain if it also has sense to say that it is not certain.

To say that there was a time when there was nothing is on the same level and as nonsensical as to say part of my visual field is not coloured.

Infinity = an infinite possibility expressed in a law.

To talk about the relation of object and sense-datum is nonsense. They are not two separate things. "This is a brown

[1] See Introduction, p. xv.

patch" and "This is a table" are two different statements having different verifications. Propositions differ according to their verifications.

Hypothesis and proposition.
    A hypothesis goes beyond immediate experience.
    A proposition does not.
    Propositions are true or false.
    Hypotheses work or don't work.

A hypothesis is a law for constructing propositions, and the propositions are instances of this law. If they are true (verified), the hypothesis works; if they are not true, the hypothesis does not work. Or we may say that a hypothesis constructs expectations which are expressed in propositions and can be verified or falsified. The same words may express a proposition to me, to you a hypothesis.

Any facts necessary to give a proposition sense enter into the propositional symbol. If a fact is presupposed in the significance of a symbol, then it is part of the symbol.

We cannot say "this" is simultaneous with "that" when "this" already contains time.

Whenever we try to talk about the essence of the world we talk nonsense.

A thought does not contain its own fulfilment. Thought, proposition (written or spoken) and plan are all on precisely the same level.

Logic may mean two things: (1) a logical calculus as e.g. the *Principia Mathematica* (2) the philosophy of logic.

We explain a rule of interpretation by giving instances of it.

Philosophy, by clarifying, stops us asking illegitimate questions.

The function of words is in no sense to *remind* us of objects.

Explanation *by means of* language and explanation *of* language are entirely different.

A proposition or plan does not contain its rule of application.

Natural laws can be justified; rules of grammar cannot be justified.

We are inclined to say that the greater the number of instances we have which conform to a hypothesis the more "probable" the hypothesis is. What we mean is the more *inconvenient* it would be to abandon the hypothesis and frame another.

Whenever the propositions constructed by a hypothesis are "always true", that is, admit of no falsification or verification, the hypothesis is merely senseless (an "idle running wheel") because neither confirmation nor lack of confirmation is possible. Eddington says that whenever you turn a light-ray on an electron it vanishes: I might also say that there is a white rabbit on my sofa which cannot be seen because whenever anyone looks at it it vanishes. These two propositions are on exactly the same level: both are merely senseless.

The relation between an expectation and its fulfilment is shown in language by the fact that we say "I expect p" to express the expectation and "p" to express the fulfilment.

When Russel analysed the proposition "The present King of France is not bald" he was giving a grammatical rule.

Any multiplicity not in the symbol will be in the rule applying to it.

The expression of a rule must itself be within a larger system of rules.

A copy is a copy only in reference to its intention. Mechanical reproduction in itself can't be wrong, and so differs fundamentally from intended copying. The expression of the intention can't contain the intention, for language can't explain itself. Language is space; statements divide space. Language is not *contiguous* to anything else. We cannot speak of the use of language as opposed to anything else. So in philosophy all that is not gas is grammar.

I hate Smith is not equivalent to I hate + Smith is the cause of my hatred.

All explanations of meaning are definitions. Proper names do not function differently as symbols from such words as "and". Their meaning is given by their use, by the rules applying to them.

You cannot say "seems to seem". Solipsism and behaviourism are opposites of each other.

You cannot say "I have a visual field". Visual and tactile space are distinguished by differences of grammar.

Distinguish between motive and cause. The causal theory of meaning confuses them. Motive is contained in act, but cause is not.

The description of intending and of playing the game will contain an expression of a general rule. So a general rule must be contained in our actually playing the game; but no expression of a general rule enters into our actual playing.

x	5
x . x $=$ general rule	5×5 actual squaring
x^2	25

No x and x . x enter the actual process of squaring; but there is something in common, for x

$$x . x$$
$$x^2$$

describes the action of the squaring. We must distinguish the expression of the rule from its use.

$\sim\sim p = p$ says nothing about negation. It says something about the type-sign $\sim$ So horse$=$quadruped genus equus is about the type-sign horse. A word can only function in a proposition, so negation can only occur in negating. Spoken alone a word may call up an image, but this is not its meaning. The image will be a state of affairs in the description of which the word would occur. A word only functions in a proposition. So the only way in which a man can show that he has understood an explanation is by using the term explained in a proposition correctly.

The geometry of visual space is not Euclidean, and we must distinguish the geometry of visual and of physical space.

We say we get nearer to $\sqrt{2}$ by adding further figures after the decimal point: 1.1412 - - -. This suggests that there is something we *can* get nearer to. But the analogy is a false one. What we give is a rule of accuracy: the more figures we add to 1.1412 - - - the closer will the square of the resultant figure be to 2.

Geometrical theorems as rules of accuracy.

To every rule or order that can be stated there must correspond a use of the rule or an obeying of the order. There must be a one-one correlation between statement and use.

p is nonsense simply means p is not a proposition. But there may be many different reasons *why* p is not a proposition.

Analysis in logic means giving grammatical rules.

We can't build up language.

No explanation can go beyond symbolism. To understand an explanation we must understand symbolism.

It is nonsense to ask "What would it be like if I did not understand language?"—though of course I may know what it is like not to understand *a* language.

If you understand at all you understand a proposition. So words have no significance apart from propositions.

If you understand you understand in a system: you understand as opposed to misunderstand.

In translating you have nothing but grammar to guide you.

You can't describe the application of language: you can't justify language. Unless symbolism had application it would not be symbolism at all.

Grammar is a description of language.

The weight (importance) of a grammatical discovery is the weight of the linguistic usage involved.

We use "necessary" both in propositions of logic and in propositions about the physical world because there is some analogy between them; e.g. "a proposition is necessarily true or false", "heavy weights necessarily fall". This analogy also comes in (in part at any rate) in our expressions for generality; e.g. "all weights fall", "all propositions are true or false". Compare also Russell's notation, variable and apparent variable, $p \lor \sim p$ (which in ed.2 becomes $(p). p \lor \sim p)$ and $(x)\phi x \supset \psi x$.

It is a fallacy to ask what causes my sense-data: and modern psychology commits a similar fallacy in ethical matters.

Notes dated October 1930 and made after discussions with Wittgenstein.

1.   If we say of an object O, "This is 3 ft. high", then the object is part of the symbol. But if we say $(\exists O). \phi O$ then O is not part of the symbol. But the propositional sign is still not complete; it cannot be understood of itself.

2. For suppose a symbolism in which all explanations have been given: we still cannot understand the symbolism simply by looking at it, simply by itself.

Some words however can be understood simply by a study of their occurrence in symbolism—the so-called logical constants. Words do not always *mean* or *point to* something.

3. When we explain a symbolism we can do nothing but add to our symbolism.

When we learn the meaning of a symbol the way in which we learn it is irrelevant to our future use and understanding of it. The way in which I learned my A B C and learned to read is irrelevant to my future understanding of written symbols—it is a matter of purely historical interest. But something does as it were *adhere* to the symbol in the process of my learning its meaning, and this becomes part of the symbol.

4. The proposition "This is red" presupposes the colour-space.

5. Language includes all that is necessary to give sense to symbols.

6. The fulfilment cannot give anything more than the expectation which it fulfils: otherwise how could we know that the expectation *had* been fulfilled?

7. To understand a rule of interpretation is to give instances of its application.

A rule of interpretation is exactly analogous to a rule of projection.

The rule of projection cannot be expressed except in actual projections, none of which contain the rule.

To make sure that we understand a rule of projection we imagine applications of it; but these imaginary applications are on precisely the same level as any others. They do not give the meaning of the rule in any sense that others don't. (The same applies to the imaginary fulfilment of an expectation.)

8.   Language is a calculus.

Thinking is playing the game, using the calculus.

Hence the question of the time taken by thought is important. Thought is not related to language in the same way that having a toothache and saying I have a toothache are related. Thought is the actual use of the linguistic calculus.

9.   An image may or may not be an essential part of a symbol. If I want to recognise Mr. Smith I must have some sort of image of him: but this may not always be necessary.

Then the symbols in the two cases, in which the image is necessary and in which it is not, are different symbols to which different sets of rules apply—though of course there are similarities between them.

Note dated Jan 1931

p	q	
T	T	T
F	T	T
T	F	F
F	F	T

In Russell's notation p ⊃ q is verified if p is False or if q is True. But clearly this is not what we mean when we say If p then q in normal usage. In normal usage we are talking in terms of hypotheses and not propositions, but, for example, if I say "If I strike a match (p) there is a flame (q)", we do not regard it as confirmation of "If p, then q" if I don't strike a match ($\sim$p, pF). Or take a statement giving a time-limit—"If x happens within the next half-hour, I shall be surprised". If x does not happen we do not regard this as confirmation of the statement. But in Russell's usage $\sim$ p does verify p $\supset$ q, which shows that our ordinary usage of If p then q differs from his.

In ordinary language we use propositions such as All S is P to mean There are Ss and all are P. Such propositions are hypotheses.

In Russell's notation, if there are no Ss, then all S is P and no S is P will both be true.

For   All S is P $= (x)\phi x \supset \psi x$

         No S is P $= (x)\phi x \supset \sim \psi x$

and $\sim \phi x$ verifies both propositions.

But in fact in both ψx is independent of ϕx; it is an idle running wheel and goes out as irrelevant.

$(x)\phi x \supset \psi x. \sim \phi x = \sim \phi x$

$(x)\phi x \supset \sim \psi x. \sim \phi x = \sim \phi x$

for p.q v $\sim$ q = p

Thus, if there are unicorns they bite, but there are no unicorns = there are no unicorns.

To sum up—

(1) If we use All (No) S is P in its common sense to mean There are Ss and all (none) are P; then if there are no Ss the assertion There are Ss is false and the rest of the proposition irrelevant.

(2) By All (No) S is P we normally mean, There are Ss such that - - -. Here the rules of conversion hold.

(3) $(x)\phi x \supset \psi x$ is not what we mean in ordinary language by p always implies q. Russell's usage is not the normal one, in which hypotheses are involved.

At some time during the year 1930–1931 I had a discussion with W. about the first four propositions (1 to 2.0121) of the Tractatus. *My notes run as follows.*

1. "The world is everything that is the case". This is intended to recall and correct the statement "The world is everything that there is"; the world does not consist of a catalogue of things and facts about them (like the catalogue of a show). For, 1.1, "The world is the totality of facts and not of things". What the world is is given by description and not by a list of objects. So words have no sense except in propositions, and the proposition is the unit of language.

1.12. "For the totality of facts determines both what is the case, and also all that is not the case". This is connected with the idea that there are elementary propositions, each describing an atomic fact, into which all propositions can be analysed. This is an erroneous idea. It arises from two sources. (1) Treating infinity as a number, and supposing that there can be an infinite number of propositions. (2) Statements expressing degrees of quality. This is red contradicts This is white. But the theory of elementary propositions would have to say that if p contradicts q, then p and q can be further analysed, to give e.g. r,s,t, and v,w, and $\sim$ t. The fact is self-sufficient and autonomous.

1.13. "The facts in logical space are the world". Logical space has the same meaning as grammatical space. Geometry is a kind of grammar: there is an analogy between grammar and geometry. Grammatical space includes all possibilities. "Logic treats of every possibility" 2.0121.

2.01. "An atomic fact is a combination of objects (entities, things)". Objects etc. is here used for such things as a colour, a point in visual space etc: cf. also above, A word has no sense except in a proposition. "Objects" also include relations; a proposition is not two things connected by a relation. "Thing" and "relation are on the same level. The objects hang as it were in a chain.

2.012. If you know how to use a word and understand it you must already know in what combinations it is not allowed, when it would be nonsense to use it, all its possibilities. So in logic there are no surprises; we must know all the possibilities. We discover new *facts*, not new possibilities. It has no sense to ask "Does red exist?"

INDEX

121

Index

Index